Scientific Writing for Young Astronomers
Part 1

EAS Publications Series, Volume 49, 2011

Scientific Writing for Young Astronomers

A Collection of Papers on Scientific Writing

Part 1

Edited by: *Christiaan Sterken*

EDP SCIENCES

17 avenue du Hoggar, PA de Courtabœuf, B.P. 112, 91944 Les Ulis cedex A, France

First pages of all issues in the series and full-text articles
in PDF format are available to registered users at:
http://www.eas-journal.org

Cover Figure

Computer rendering of the Herschel satellite overlaid on a PACS/SPIRE image of a stellar nursery in Aguila, where hundreds of stars are forming (top). The lower panel shows a SPIRE three-colour image of GOODS-N with thousands of visible galaxies with an inset of the famous JCMT/SCUBA image for comparison.
© A&A cover Volume 518, Part I (2010).

Indexed in: ADS, Current Contents Proceedings – Engineering & Physical Sciences, ISTP®/ISI Proceedings, ISTP/ISI CDROM Proceedings.

ISBN 978-2-7598-0506-8   EDP Sciences Les Ulis
ISSN 1633-4760
e-ISSN 1638-1963

100658898x

Table of Contents

Expanded Table of Contents

Preface

The *Astronomy & Astrophysics* (A&A) Board of Directors and EDP Sciences, the Publisher of A&A, decided in 2007 to organize a School on the various aspects of scientific writing and publishing, with two aims: (a) to teach young PhD students how to express their scientific thoughts and results through adequate and efficient written communication, and (b) to discuss the operation of A&A as an example of an international peer-reviewed journal in astronomy. To this end, a first SCIENTIFIC WRITING FOR YOUNG ASTRONOMERS (SWYA) School was organized in 2008, and a second one in 2009. The Schools were financed by A&A and by EDP Sciences, who also provided all logistic support.

This book contains a collection of articles aimed at supplying guidelines to PhD students and postdoctoral fellows to help them compose and publish scientific articles. Several articles present examples and case studies specially prepared for these Schools. The organization of this Volume, though not a faithful record of the lectures, closely reflects the structure of the Schools, and contains articles about the editorial office and the Editors, the Publisher, the business models in journal publishing, the journal *Astronomy & Astrophysics*, the refereeing process, bibliometric databases, and library services. The second Volume deals with the writing process, communication with graphics, bibliometry and ethics.

The first three articles of this Volume are presented by Editors – past and present – of A&A. The first article starts with a brief historical introduction, followed by a review of the scope, editorial process, and production organization of A&A, including the economic model of the Journal, and some current issues in scientific publishing. This article is immediately followed by a paper on the history of A&A, by one one of the first Editors-in-Chief of the Journal. Then follows a chapter that explains the functioning of scientific journals from the editorial side of the process, with emphasis on the interaction between the three parties: the author(s), the Editors, and the referees. The chapter explains how manuscript evaluation proceeds, its advantages, pitfalls, and criteria – scientific, archival, and ethical – and how the procedures have evolved historically and consensually.

Then follows an article on the role of the Publisher, describing the various facets of the publishing activity, including the daily work of a dedicated team of professionals that guarantees the timely production of this international weekly astronomy journal.

Two articles deal with language editing: the first reviews the general advantages of editing the English expression, and describes both the aims of this effort and its place in the full publication process. That article is followed by a detailed Guide to language editing that serves as a reference for any non-native user of the English language.

The remainder of this Volume is dedicated to technical support in the broadest sense, portrays astronomy libraries as a tool, and describes the support by the databases at the *Centre de Données astronomiques de Strasbourg* (CDS). Special attention is given to the two most important information tools in astronomy: the *NASA Astrophysics Data System* (ADS) and the arXiv e-print server, in addition to an introduction to bibliometric studies. The last article describes how published data find their way to the CDS databases, and focuses on the strategy that ensures a high level of quality.

The Schools took place in Hotel and Conference Center Aazaert in the city of Blankenberge on the North Sea coast of Belgium. The meetings were very rewarding, and happened in a very pleasant atmosphere, thanks to the dedication and professionalism of the complete staff: from the barkeeper to the waiters, and from the administrative staff to the chefs in the kitchen – they all contributed greatly towards making us feel at home and have an unforgettable time. In the name of all participants, I express deep appreciation to Mr. Peter Kamoen, Manager of Hotel Aazaert, and to his entire staff.

The Manager of the Schools (on behalf of EDP) was Mrs. Isabelle Turpin, assisted by Mrs. Aurélie Chastaingt. They did a professional job preparing each event, and their presence during the duration of each School was a blessing for anyone who needed help. In the name of all lecturers and students, I extend a warm thanks to Isabelle and to Aurélie.

Dr. Agnès Henri, Mrs. Sophie Hosotte, Mrs. Isabelle Houlbert and Mrs. Marie-Louise Chaix (EDP) are given thanks for their efficient communication and helpful support during the preparation of this Volume.

Christiaan Sterken
Editor
University of Brussels, Belgium

DOI: 10.1051/eas/1149000

About the Authors

Joli Adams is a graduate of Cornell University and UCLA. She has taught English language, literature and writing in French international lycees and colleges, has edited English articles in Europe since 1983, and is now living in Paris, where she is one of the English-language editors at *Astronomy & Astrophysics*.

Claude Bertout, Editor in Chief of *Astronomy & Astrophysics* since 1999, is a research professor at the Paris Astrophysical Institute. His main research interests are the physics of young solar-type stars and their circumstellar, protoplanetary disks.

Thomas Boch is an engineer at the *Centre de Données Astronomiques de Strasbourg*, in charge of the CDS portal and, among other things, one of the *Aladin* developers.

Laurent Cambrésy is an astronomer at the *Centre de Données Astronomiques de Strasbourg* (CDS). He provides science expertise for SIMBAD and VizieR databases. His main research topics are the interstellar medium and star formation in nearby molecular clouds.

Thierry Forveille is an astronomer at the *Observatoire de Grenoble* and one of the Associate Editors of *Astronomy & Astrophysics*. His main field of current research is observational studies of brown dwarfs and extrasolar planets.

Françoise Genova has been the Director of the *Centre de Données Astronomiques de Strasbourg* since 1995. She has led the transition of the data center into the WWW era and has been involved in the development of the international astronomical *Virtual Observatory* from the very beginning.

Uta Grothkopf is Head Librarian of the *European Southern Observatory* (ESO) library in Garching near Munich, Germany. The library fulfills the information needs of ESO astronomers and engineers and develops digital services to give greater access to scientific content. The library also maintains the Telescope Bibliography, a database of refereed articles that use ESO data, and provides ESO Management with bibliometric statistics derived from this database.

Claire Halliday has a PhD in extragalactic astronomy from the University of Durham and eight years of postdoctoral research experience in this research field. She has experience of writing and publishing astronomical articles and has been a language editor at *Astronomy & Astrophysics* since January 2008.

Agnès Henri is publishing director at EDP Sciences, the Publisher of *Astronomy & Astrophysics*. She has a PhD in physics from Université Paris VII-Diderot.

Cécile Loup is an astronomer at the *Centre de Données Astronomiques de Strasbourg* and coordinator for the science aspects of *SIMBAD*. Her main research topic is evolved stars.

François Ochsenbein is an astronomer at the *Centre de Données Astronomiques de Strasbourg*, in charge of the catalogs and the VizieR database.

Astrid Peter holds a Masters degree in English from the Georg August University in Göttingen and is a certified scientific journalist from the Berlin Journalistenschule. She studied in St. Andrews, Scotland, and Göttingen and lived in Boulder, CO, for several years. She has been a language editor at *Astronomy & Astrophysics* since 2009 and also works freelance as a translator and editor for German and English.

Stuart Pottasch, professor at the Groningen Observatory, was Editor in Chief of *Astronomy & Astrophysics* from 1968 to 1976. He then became A&A Letters Editor and kept this position till he retired in 1997. His main research interests are the physics of planetary nebulae and the late stages of stellar evolution.

Steve Shore is an Associate Editor of *Astronomy & Astrophysics*, professor of astrophysics in the *Department of Physics "Enrico Fermi"* of the University of Pisa, and an associate of the INFN-Pisa and INAF-Arcetri. His principal areas of interest are theoretical and observational: hydrodynamics (especially turbulent processes), spectroscopy, radiative transfer, and recently high-energy (GeV, TeV) and gravitational-wave astronomy.

Martine Usdin is a native English speaker with a PhD in genetics from Stanford University. She has worked in research and in scientific writing. She has been a language editor at *Astronomy & Astrophysics* from 2001 to 2009.

Marc Wenger is an engineer at the *Centre de Données Astronomiques de Strasbourg*, where he is in charge of the *SIMBAD* database.

Acronyms

A&A	Astronomy & Astrophysics
AAS	American Astronomical Society
AAAS	American Association for the Advancement of Science
AAUP	American Association of University Professors
ACS	American Chemical Society
AD	Anno Domini
ADS	NASA Astrophysics Data System
AJ	Astronomical Journal
ALPSP	Association of Learned and Professional Society Publishers
APS	American Physical Society
ASP	Astronomical Society of the Pacific
ApJ	Astrophysical Journal
ApJS	Astrophysical Journal Supplement Series
BC	Before Christ
BSI	Bibliographic Star Index
CC	Creative Commons
CDS	Centre de Données astronomiques de Strasbourg
CERN	Organisation Européenne pour la Recherche Nucléaire
Chandra	NASA Chandra X-ray Observatory
CSI	Catalogue of Stellar Identifications
CV	Curriculum Vitae
DESY	Deutsches Elektronen-Synchrotron
DOAJ	Directory of Open Access Journals
DOI	Digital Object Identifier
DSS	Digital Sky Survey
EDP	EDP Sciences
ESA	European Space Agency
ESO	European Southern Observatory
FNAL	Fermi National Accelerator Laboratory, Fermilab
GS	Google Scholar
HEASARC	High Energy Astrophysics Science Archive Research
HIPs	High-Impact Papers
HST	Hubble Space Telescope
HyperLeda	Lyon-Meudon extralactique database
IAU	International Astronomical Union
ICRS	International Celestial Reference System
IF	Journal Impact Factor
INSEAD	Institut Européen d'Administration des Affaires
INSU	Institut National des Sciences de l'Univers

IOP	Institute of Physics
IOPP	Institute of Physics Publishing
IPAC	Infrared Processing and Analysis Center
JCR	Journal Citation Report
LANL	Los Alamos National Laboratory
LISA	Library and Information Services in Astronomy
LMC	Large Magellanic Cloud
MAST	Multimission Archive at STScI
MMS	Manuscript Management System
MNRAS	Monthly Notices of the Royal Astronomical Society
MSFC	Marshall Space Flight Center
MSX	Midcourse Space Experiment
NAS	National Academy of Science
NASA	National Aeronautics and Space Administration
NED	NASA/IPAC Extragalactic Database
NOMAD	Naval Observatory Merged Astrometric Dataset
NRAO	National Radio Astronomy Observatory
NVSS	NRAO VLA Sky Survey
OA	Open Access
OAI-PMH	Open Archives Initiative Protocol for Metadata Harvesting
OpenDOAR	Directory of Open Access Repositories
ORCID	Open Researcher & Contributor ID
OWL	Online Writing Laboratory
PASP	Publications of the Astronomical Society of the Pacific
RAS	Royal Astronomical Society
RSS	Rich Site Summary
SAO	Smithsonian Astrophysical Observatory
SDSS	Sloan Digital Sky Survey
SI	International System of Units
SIMBAD	Set of Identifications, Measurements, and Bibliography for Astronomical Data
SPIRES-HEP	High Energy Physics database, Stanford
STScI	Space Telescope Science Institute
UCD	Unified Content Descriptor
UNESCO	United Nations Educational, Scientific & and Cultural Organizatic
URL	Uniform Resource Locator
USNO	United States Naval Observatory
VLT	Very Large Telescope
WG Lib	IAU Comm. 5 Working Group Libraries
WMAP	Wilkinson Microwave Anisotropy Probe
WoS	Web of Science
XMM	X-ray Observatory
ZAMS	Zero-age main sequence

Scientific Writing for Young Astronomers
C. Sterken (ed)
EAS Publications Series, **49** *(2011) 1–22*
www.eas-journal.org

ASTRONOMY & ASTROPHYSICS: AN INTERNATIONAL JOURNAL

C. Bertout[1]

Abstract. After a brief historical introduction, we review the scope, editorial process, and production organization of A&A, one of the leading journals worldwide dedicated to publishing the results of astrophysical research. We then briefly discuss the economic model of the Journal and some current issues in scientific publishing.

1 Introduction

European astronomers first contemplated the possibility of establishing an international journal devoted to publishing the results of their research in the wake of ESO's creation. The preliminary discussions that eventually led to establishing CERN and ESO started less than ten years after the end of World War II. The main protagonists of the war were only beginning to emerge from the darkest hours and scientific research was at the top of the development priorities of most countries. At that time, the resentment of former enemies was still very strong in the general populations of all countries involved in the war, but many intellectuals and politicians were pushing common European projects as a way to counteract the nationalistic mind-set that had led to the two world wars. UNESCO played a decisive role in promoting scientific cooperation among Europeans; CERN was born in 1954, and ESO followed ten years later.

In Europe, astronomy was lagging far behind American research at that time, not only because three successive wars had ravaged Europe, but also because the USA had been able to build large telescopes early in the 20th century while none of the largest European countries, engaged in endless military build-up, as well as civil destruction and reconstruction cycles, had the means to develop large instruments. As a measure of the abyss between the two continents, one only need recall here that the 100 inch telescope at Mount Wilson saw its first light in November 1917 while the 193 cm Haute Provence Observatory telescope, which remained Europe's largest optical instrument until well into the 1970s, started operation only in 1958.

---

[1] A&A Editor in Chief, Observatoire de Paris, 61 Av. de l'Observatoire, 75014 Paris, France

©The authors, published by EAS, EDP Sciences, 2011
DOI: 10.1051/eas/1149001

The publishing situation in astronomy also reflected Europe's divisions. There were a number of national journals devoted to publishing national research in their country's own language, as well as several observatory publications where the observational data were archived. By the end of the 1950s, frustration was running high among European astronomers because their work had very little impact outside their own countries. In stark contrast, American astronomers had established the ApJ and AJ already by the end of the 19th century as the main vectors of communication for the entire US community. On this front, European astronomy was also lagging more than fifty years behind the US at the end of the second world war.

Stuart Pottasch, who with Jean-Louis Steinberg was a major proponent of the merging of various national astronomy journals into a single international one, has summarized the early history of A&A in Pottasch (1999). For completeness, this article is included as the next chapter in this book. As Pottasch clearly states, the new journal was to be run by astronomers for astronomers, and to this day, forty-two years later, A&A has remained faithful to this objective.

A&A in fact belongs to the astronomers of the countries that sponsor its operations through modest annual contributions of the national agencies that fund astronomy. The sponsoring countries form the A&A consortium. They subsidize and run the Journal's operations, in return for which authors from these countries are not charged for publishing in A&A. Each of the sponsoring countries elects one representative to the A&A Board of Directors, the owning and governing body of the Journal. The Directors in turn appoint the A&A Editors, who are responsible for the contents of the Journal and are accountable to the Board.

A&A is a wide-ranging journal: it publishes all new results of astronomical research, regardless of the technique used to obtain them, and accepts papers from any origin, i.e., it is not restricted to serving the sponsoring countries of the Journal. Papers published in A&A should present new astronomical results or ideas of sufficient interest to the community as concisely as possible while being comprehensive enough to allow for the duplication of the results by others.

Over the years, A&A has grown to become one of the largest astronomy journals (currently more than 20,000 pages per year) and is serving researchers from about sixty countries. Twenty-four of them currently sponsor the Journal; these are not only European but also South American. The current sponsors are Argentina, Austria, Belgium, Brazil, Bulgaria, Chile, Croatia, the Czech Republic, Denmark, Estonia, Finland, France, Germany, Greece, Hungary, Italy, Lithuania, the Netherlands, Poland, Portugal, the Slovak Republic, Spain, Sweden, and Switzerland.

2   The scope and ethical requirements of A&A

A&A is a wide-ranging journal: it publishes all new results of astronomical research, re-gardless of the technique used to obtain them, and accepts papers from any origin; i.e., it is not restricted to serving the sponsoring countries of the Journal. Papers published in A&A should present new astronomical results or ideas of sufficient interest to the community as concisely as possible, while being comprehensive enough to allow for the duplication of the results by others.

Like other refereed journals, A&A requires its authors to adhere to the strict ethical principles that rule academic endeavors. Respect of intellectual property is the most important of these principles. The copyright on publications represents one way to protect intellectual property, and academic journals usually request that authors transfer the copyright of their works to them. The copyright holder of A&A articles is ESO, which in turn transfers the copyright for a limited amount of time to the A&A publisher. In the case of A&A, the transfer of copyright does not keep the author from advertising her/his work in the most efficient way (*e.g.*, by distributing it widely in abstract and reprint databases). When speaking of copyright we are touching on much wider issues such as the so-called open access to scientific information. We will come back to these in Section 5.2. We only need to stress here that the respect of intellectual property has several practical consequences for authors, and we mention them in turn below, beginning with plagiarism.

Plagiarism is defined in the Instructions for Authors as *"the act of reproducing text or other content from works written by others without giving proper credit to the source of that content"*. Citing a copied text literally is not the only condition for determining plagiarism, which also includes any paraphrased text that discusses an already published idea without citing its source[1].

Plagiarism is a major ethical breach and may also constitute a legal breach of copyright if the reproduced material has already been published. This is particularly true when authors cite text from their own previously published works. A&A Editors refer to this as *"self-plagiarism"*. Authors who wish to quote directly from other published work must cite the original reference and include any cited text in quotation marks. Figures may only be reproduced with permission. Because A&A focuses on publishing results of original research, authors are discouraged from using any direct quotations and figures[2] of previously published papers. Software tools are available to help Editors locate plagiarized material in submitted articles and are used routinely at A&A.

Papers published in A&A should cite previously published papers that are directly relevant to the results being presented or discussed. *Improper attribution – i.e.*, the deliberate refusal to cite prior, corroborating, or contradicting results – represents an ethical breach comparable to plagiarism.

Plagiarism, self-plagiarism, and improper attribution can result in the summary rejection of a manuscript submitted to A&A. In the severest cases of plagiarism, offending authors can be banned from publishing in A&A for a determined period of time. In such cases, the Editor in Chief can also inform the Editors in Chief of the other professional

---

[1]Reproducing or adapting in one's writings material taken from Internet pages, *e.g.*, from Wikipedia articles, without mentioning the original source also constitutes plagiarism. The fact that Wikipedia is a free encyclopedia to which everyone can contribute does not mean that its contents can be used freely. Wikipedia's terms of use state clearly that *"Each copy or modified version that you distribute must include a licensing notice stating that the work is released under the Creative Commons Attribution/Share-Alike License 3.0 (Unported) and either (a) a hyperlink or URL to the text of the license or (b) a copy of the license.* Similar restrictions hold for most material available on the Internet.

[2]Figures that have appeared elsewhere (*i.e.*, copyrighted pictures) are not acceptable unless they are absolutely necessary for understanding the article submitted to A&A. In such a case, the author should request written permission of reproduction to both the original publisher of the figure and its author and send these documents to the editorial office when submitting the article.

astronomy journals of this ethical misconduct. Mercifully, such breaches remain a relatively rare occurrence.

When you submit an article to A&A, you attest that your coauthors have read the work and agree with the paper's contents. Adding prominent scientists to the authorship, although they had nothing to do with the paper and do not know about it, is a practice that seems to be developing in some countries, perhaps in the hope that the article will be accepted more easily. This is, of course, another reason for summary rejection.

One should also mention that any given paper cannot be submitted simultaneously to more than one refereed journal. We must emphasize in this context that A&A does not systematically refuse papers that have been previously rejected by another journal. However, the Editors expect a frank attitude from authors concerning these papers. If an author tells the Editor about that paper's previous history, it is likely that A&A will give it a second chance to be published. Some authors choose not to tell us when their submitted paper has already been refused elsewhere, but since we use the same pool of referees as other leading journals, chances are that the Editor in charge of the paper will find out anyway. In this case, summary rejection is likely to follow.

Finally, a given work cannot be published in more than one refereed journal. Doing so – and we have seen a couple of cases in the recent past – is *not* the best idea to increase the length of one's list of publication. Instead, it constitutes the ultimate ethical and copyright breach, and results in a long-term ban of the culprit from publishing in A&A.

3   The A&A community

In 2010, we received about 2400 articles and Letters to the Editor from 62 different countries, and 77% of these manuscripts originated from sponsoring countries. As shown in Figure 1, the main individual countries contributing to the contents of A&A are Germany (which includes ESO), France, Italy, the United Kingdom and Spain, and the Netherlands. Among these top contributors, the United Kingdom is the only one that does not sponsor A&A. Figure 2 illustrates the wide dispersion in the number of submissions received from A&A sponsors, in good part reflecting the number of astronomers in the respective countries.

In contrast to other astronomy journals, A&A uses sections to order the published papers. Each section corresponds to a broad scientific area of astronomy or astrophysics. This helps the reader to more easily find the papers that match her or his scientific interest in the table of contents. As it turned out, the division in sections allows for a natural and progressive transition to online-only publication of the Journal. We shall come back to this issue later.

The percentage of papers received in 2010 in each of the scientific sections are shown in Figure 3: 19% of the papers we receive concern stellar physics, 16% discuss extragalactic astronomy results, and 13% deal with interstellar medium studies. This distribution has remained fairly constant over the last decade. We have seen a steady increase in the submissions in cosmology and extragalactic astronomy. More recently, we have been receiving increasing numbers of papers concerning planets and planetary systems, which mainly reflects the current advances in extrasolar planet searches and in planet formation

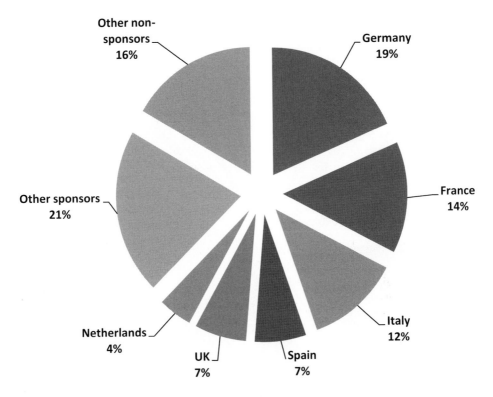

Fig. 1. Origin of A&A submissions in 2010.

theory. This evolution reflects the driving role of the ESO VLT in European astronomy today and of community access to several successful European space and ground observatories from X-ray to millimeter wavelengths.

4   The A&A peer-review process

The peer-review process of submitted articles and Letters to the Editor is done in several steps described briefly in the following[3]. Steve Shore's chapter in this volume provides additional information on the peer-review process as seen from the editorial side.

4.1   First step: Editor's initial reading

The Editors of A&A are responsible for the contents of the Journal on behalf of the Board of Directors. This responsibility can only partially be shared with referees since it involves issues both of Journal policy (Editors enforcing decisions made by the Board of Directors)

---

[3]This section is an abridged and updated version of a previously published editorial by Bertout &Schneider (2004).

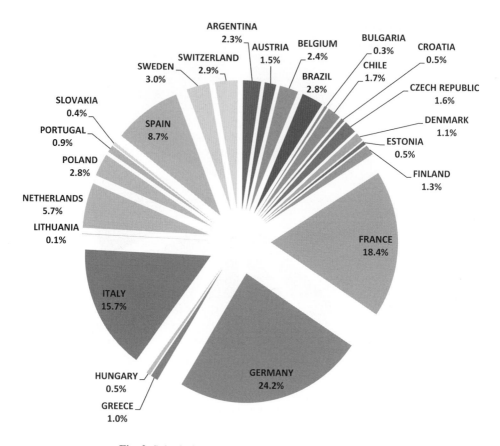

Fig. 2. Submissions from sponsoring countries in 2010.

and of scientific excellence, for which the Editors rely mainly on the expertise of referees. Deciding whether a submitted paper should be sent to a referee for a scientific evaluation is therefore the first question that the Editors ask themselves upon receipt of a new regular paper or Letter.

4.1.1   Regular articles

In their initial reading of a new submission, the Editors are not judging the scientific value of the paper in detail (this is the referee's role), but must gauge its potential scientific interest and information content, because both are important aspects of new submissions. A paper presenting an important result of high scientific interest can see its information content diluted by details or digressions that are unnecessary for understanding the result. Conversely, a catalog with large information content might have little *apparent* scientific interest but will be extremely useful to the community for further astrophysical research.

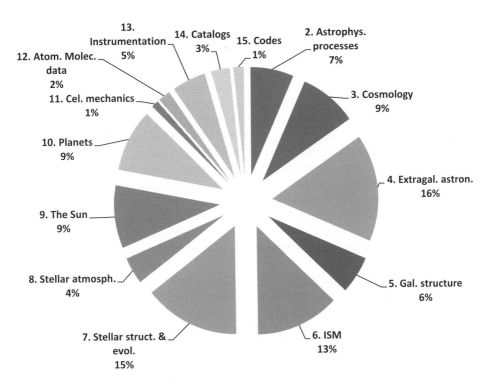

Fig. 3. Percentage of papers received in 2010 in each scientific section.

The information content required of new submissions is something that has varied with time. Forty years ago, when astronomical information was much scarcer than now, it was easy for researchers to publish at least one paper after every short observational run on a 1 to 2 m-class telescope, because there were indeed new and useful ideas to be gained even from limited observations of exotic celestial objects. Today, the situation is entirely different, thanks to the huge amount of astronomical information that flows from large optical telescopes and radio arrays, space observatories, and surveying telescopes. As a consequence, the typical information content of a paper must be substantially greater for the journals and readers to keep up with the information flow and stay up-to-date with the science that follows from it.

We should also mention here that A&A is result-oriented; *i.e.*, Editors will not consider a paper presenting, say, an unsuccessful search for some exotic celestial objects *unless* this null result has some interesting astrophysical implications, which are equivalent to exciting results and likely to lead to further science.

Approximately 95% of all the articles that we receive are sent on to referees, but the reasons given above explain why an A&A Editor might either not do so or might request some changes in the paper prior to sending it to a reviewer.

4.1.2   Letters to the Editor

In addition to the criteria described above, Letters should require rapid publication, as written in the *Instructions for Authors*. This criterion is not very well defined, though, and its meaning has changed with time. There are now other ways to communicate essential information quickly than in journals. In particular, we note a clear shift towards using the *astro-ph* preprint server as a communication channel for timely results in the extragalactic community. In addition, publication times in the Main Journal are not much longer than for the Letters. What, then, makes a manuscript a Letter?

To qualify for a Letter, a manuscript should be both timely and of very high quality. In addition, the manuscript should not exceed 4 journal pages. Automatically, then, manuscripts are not considered for publication as a Letter if their contents appear less than timely – for example, if the reference list contains no paper on the subject from the past few years – or if it is too long. The length restriction has been relaxed at times, but a manuscript with 5 pages must be exceptional, and referees are given particular instructions about this.

In contrast, it is against editorial policy to sacrifice clarity to brevity. Trying to shorten a manuscript to Letter length at the cost of legibility is therefore not acceptable. It is also against editorial policy to defer essential information to a later, longer manuscript. An expert reader must be able to judge a scientific paper based on the information presented in the work or already available in the literature. A reference to work "in preparation" is not useful for evaluating a paper's contents and the manuscript may either need expansion – and thus most likely exceed the page limit for a Letter – or be postponed until the article referred to becomes publicly available.

4.1.3   Articles on special subjects

Although the Instructions for Authors state that *"A&A publishes papers on all aspects of astronomy and astrophysics"*, there are topics that are on the borderline of the subjects covered by A&A and the other main astronomical journals. Examples include the theory of strange stars, the physics of gravitational waves, some aspects of celestial mechanics.

One quick way we check how well a subject is covered in the major astronomical journals is to look at the paper's reference list. If none, or hardly any, of the cited papers have appeared in the major astronomical journals, then the Editor needs to consider the question of whether this paper should be published in A&A and not in one of the specialized journals to which the manuscript refers. No strict policy for these cases has been formulated yet, however, and decisions are made on a case-by-case basis. Referring to the example of strange stars, such a manuscript can be published in A&A if, *e.g.*, predictions are made that can be tested astronomically, say by properties of pulsars, but will normally be rejected if it does not discuss astrophysical consequences. Similarly, a paper devoted to some theoretical issue in celestial mechanics that is not applied to an astronomical object will usually be rejected.

One criterion to be applied here is whether this article will interest people who regularly read A&A. In these cases, the Editor usually cannot rely on advice from the referee alone, as the referee is an expert on the subject and will therefore almost certainly find it relevant. Furthermore, the criterion will change in time: whereas at present neutrino experiments are usually outside the range of topics considered for A&A, future experiments will open the window to neutrino astronomy, which then clearly becomes a topic of interest to the A&A readership. A similar situation is expected to develop for gravitational waves.

4.1.4   Some style issues

Editors may momentarily refuse to consider a paper not for scientific reasons, but because its style and language are obstacles to understanding the science. While linguistic and stylistic improvements can be made at the editorial office upon acceptance of the paper, we are not in a position to rewrite badly written papers before sending them on to referees, nor should we be. We used to seek referee reports for all but the most poorly written papers, but referees often complained. We also noticed that the peer-review process of those papers was less constructive and took a much longer time than average. We are thus now making certain before starting the peer-review process that the language is not an obstacle to comprehending the science. While most A&A authors are non-native English speakers who cannot be expected to write perfect English, those who have not yet developed full control of the language are urged to find a native English-speaking colleague or friend to read and correct their work.

The *Instructions for Authors* provide precise guidelines concerning the style that Editors expect, and new contributors to the Journal are kindly urged to read them carefully before writing their first papers. The manuscript preparation as a LaTeX file, now standard in all major journals, has put an additional burden on authors, because they are expected to prepare an almost print-ready version of their paper. However, besides a burden, this also provides the author with nearly full control over the manuscript and leaves much less room for mistakes during the final processing of the paper at the publishers. To exercise this control, however, the style guidelines need to be followed exactly.

Two areas where we encounter many style difficulties are the table titles and figure captions, which are often much too long although the instructions to authors clearly state that *"the scientific discussion of the table or figure contents should appear in the main body of the article, not in the table title or figure legend"*. Please always refer to the *Instructions for Authors* for the guidelines to be followed when preparing tables and figures.

A few words about the length of papers is now in order. Over many years, the average length of A&A papers has been constant at 10 printed pages. In the last several years, we have witnessed an increase in the papers' average length that has been partly offset by changes in the layout of the articles (less spaces between characters and lines, smaller figures, smaller fonts for references, etc.). While the scientific content of many papers fully justifies their length, a number of submissions appear verbose and unfocused. Since A&A is an international journal, contributors need to cultivate the usual unadorned style of international scientific communications (*cf.* Joli Adams's chapter on English language editing at A&A).

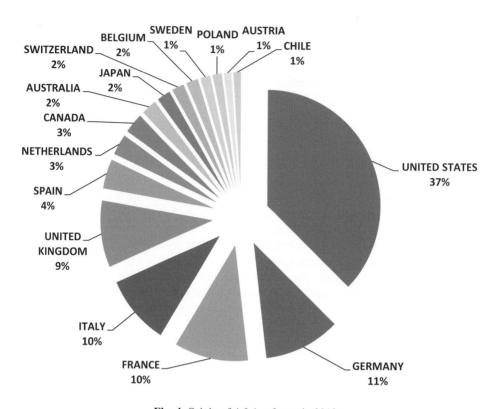

Fig. 4. Origin of A&A referees in 2010.

4.2 Scientific evaluation by the referee

Once the Editor concludes that a newly submitted paper falls within the scope of A&A and is reasonably well organized, a referee is chosen to review and evaluate the scientific contents of the paper. Editors look for the best possible referee for any given paper without resorting to pre-existing lists[4] but by using ADS to find a potential referee within the group of people who have worked on the topic of the submitted paper in the recent past. If the field of research and/or the people involved are unfamiliar to the Editor, she or he looks at the recently published works to find out whether there are competing groups on the same topic, who the most cited people are, and so on. With the online tools available nowadays and some experience, the Editor can quickly get a rough idea on any astronomical research field, which is usually sufficient to make a good choice of referee.

Next, we should point out that we do not restrict the choice of referees to those scientists who work in A&A sponsoring countries. Because astronomical research exists in a framework of global cooperation, the peer-review process also needs to be global. This furthermore ensures that the scientific criteria for paper acceptance are comparable at the

---

[4]We do, however, keep a master list of referees that includes how often they have been called by A&A Editors.

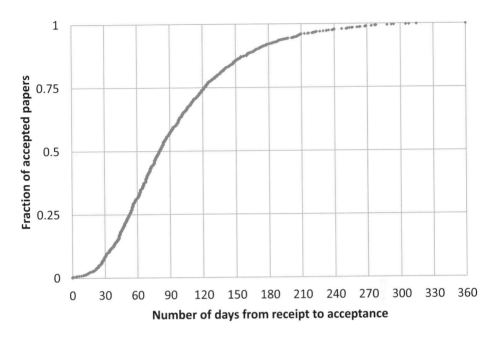

Fig. 5. Cumulative distribution of acceptance times for A&A papers in 2010.

main astronomy journals, as demonstrated by their comparable rejection rates. Figure 4 shows the origin of reviewers who evaluated A&A papers in 2010. Only countries contributing more than 1% of all referees are shown there. We note in particular that 40% of all A&A referees are from North America, where a large fraction of the global astronomical community is working; conversely, the US journals call on many European referees.

We usually use only one referee for any given paper, except when a paper deals with more than one subfield of astronomy, in which case we request the assistance of a specialist in each subfield. This is, however, a rare occurrence. We also usually request a second referee opinion in case of conflict between author and first referee. A few years ago, we tested using two referees for randomly selected papers for a short time. This experience did not convince us that the peer-review system was improved much by adding a referee.

In some cases, finding a potential referee turns out to be difficult. Examples here are papers in a highly specialized field. Another example are papers authored by a large group in a community, such as happens when new facilities go into operation; at this point, it is difficult to find an outside person competent enough to judge the technical aspects of a manuscript. In such cases, we sometimes ask the authors to provide us with a list of a few competent, independent potential referees, and most of them are quite willing to come up with such a list. After a quick check with ADS, appropriate referees are then identified.

Referees focus primarily on the scientific contents of the paper. A&A provides a short questionnaire as guidelines to the various aspects of the evaluation that the Editors would

Fig. 6. Cumulative distribution of acceptance times for A&A Letters in 2010.

like the referee to address. A report prepared by a seasoned referee will not necessarily address all these questions explicitly but does so implicitly instead. This is perfectly acceptable, as the questionnaire is mainly intended to help referees with little previous experience of the peer-review process.

After more than ten years of experience as Editor-in-Chief, I can testify that the peer-review process works extremely well in the vast majority of cases. Many colleagues consider that refereeing is an integral part of their professional duties and are keen to deliver constructive and thoughtful reports within a reasonable time frame. Authors revise their works in accordance with the referee's requests and suggestions. In most cases, two iterations with the referee are needed before the referee recommends acceptance of a submitted manuscript. The process takes three months on average, but a more meaningful indicator of the time needed for acceptance of a submitted paper is shown in Figure 5, where the fraction of papers accepted in 2010 is displayed as a function of the time between submission and acceptance. About 10% of the papers – presumably the best ones – are accepted within one month, and the median acceptance time is 81 days. Three-thirds of all papers are accepted within four months but the tail of the distribution extends to one year.

These times include the times needed by authors for revising their paper. The Editors once *assumed* that this was as short as could be so that there was no need to question it further. Consequently, they attributed most of the delay in the time from paper submission to acceptance to the referee. This assumption is in fact incorrect and very unjust to the referees. With the web-based manuscript management system in place, the Editors now know how many papers are in revision at any given time and, to their surprise, noticed that there are many more papers being revised than there are papers with a referee (the ratio is approximately two to one). This is an apparent indication that on average *the time spent in revising an article is twice as long as the refereeing time.*

This issue is particularly relevant for the Letters, in particular when authors try to argue about the urgency of the publication of a manuscript, even though it took them several months to prepare a revised version following the referee report. In these cases, manuscripts are more likely to be forwarded to the Main Journal, since apparently the authors are not all that convinced of the time factor themselves. Figure 6 displays the cumulative distribution of acceptance times for Letters in 2010 and shows that 36% of the submitted Letters are accepted within one month from receipt, 50% within 33 days, and 90% are accepted within two months. As expected, this is significantly faster than acceptance times for regular papers. We also note that 29% of submitted Letters were transferred to the Main Journal in the course of peer review, usually because the Editor or referee is not convinced of the urgency of publication.

We should mention here that the global acceptance rate of A&A articles is about 80%. There are strong deviations from the mean among the different scientific areas. For example, the rejection rate of articles submitted for the "Astrophysical Processes" section of the Journal is about 40%, because a number of manuscripts received for this section are the so-called "crackpots"[5] that are unsuitable for publication in any scientific journal. The main sections of the Journal (extragalactic astronomy, interstellar and circumstellar studies, structure and edvolution of stars, stellar atmospheres, the Sun) have rejection rates ranging from 8 to 17%. Also, papers originating from sponsoring countries are rejected less often (14% rejections) than papers originating from non-sponsoring countries (40% rejections). This is *not* because Editors favor papers submitted by sponsoring authors. Quite the contrary, *all papers go through the same peer-review process regardless of origin.* As a proof of this policy, let us state that one of the countries with the lowest rejection rate is a non-sponsoring country: the UK, with only 7% rejections. Instead, the reason is simply that some of the non-sponsoring countries that contribute most to A&A submit a number of manuscripts of dubious scientific interest.

4.3 Complications in the peer-review process

Of course, there are also several possible causes of complications in the course of refereeing that we discuss below in some detail. When major problems arise, the A&A Editor in Chief is likely to intervene in the process and help the various parties involved in finding a reasonable solution.

Besides the ethical issues that may result in summary rejection from submitted papers (*cf.* Sect. 2, other less serious but more frequent problems can occur during the peer review are less serious but more frequent. They are usually related to communication

---

[5] As an example, the commemoration of Albert Einstein's achievements in 2005 brought to us a flurry of papers aimed at disproving general relativity on the grounds of a variety of flawed arguments.

delays between the different actors of the scientific validation that result in delaying the decision of whether to publish the submitted papers.

4.3.1   Delay in finding a referee

Most senior astronomers are asked by the various astronomy journals to provide at least one referee review every year, and the most well-known of them are asked to provide several reviews. This strong demand on referee time has the obvious consequence that referees may refuse to review a paper on the grounds either of being already busy with a paper for another journal or of having done several reports recently. However, A&A Editors have noticed that referees seldom refuse to review a paper when they really want to know what it contains, so this appears a primary incentive for a scientist to accept a refereeing request. Providing the paper's abstract, as A&A does routinely, also helps referees decide whether they can competently deal with the paper. Nevertheless, one senses that the fraction of refereeing requests that are refused is slowly increasing over the years. One reason for this is the growing pressure on the time of senior scientists. In addition to the demand on refereeing for journal articles, senior researchers are faced with an increasing number of reviewing and report requests from funding and evaluation agencies.

In many cases, the Editor will quickly think of someone who would be eager to see the submitted paper so there will be no need to ask several people in turn, a process that usually results in long delays. In recent years, the median time between the time of submission registration and the time at which a regular article goes to the referee has been 7 to 8 days. However, the tail of the distribution extends to more than thirty days. For the Letters, these times are somewhat shorter; nevertheless, there are exceptional cases where it takes more than three weeks before a referee has been located who is willing to review a manuscript.

4.3.2   Delay in getting the referee report

As already mentioned, competent referees are often extremely busy with a variety of other duties besides their own research and/or teaching. The likelihood that the review request will end up being postponed or even forgotten by the referee is therefore not negligible. A&A Editors usually send a first reminder to the referee three to four weeks (two weeks for Letters) after the task was accepted, and kindly request an answer. The worst-case scenario is when a referee answers the first reminder by saying that the report is under way, but after two or three reminders, does not bother to answer anymore and does not send a report. The Editor then needs to start the process over and find another referee. This is when the longest delays occur. There is not much that the Editor can do to reduce this delay, except to explain the situation to the second referee and kindly request a report on a very short time-scale. In such cases, the Editor will often contact trusted personal friends and colleagues who are kept "in reserve" for dealing with these emergencies.

Often, Editors will be contacted by authors who believe that the referee is actually trying to delay the paper because of supposed competition with the authors. We would like to emphasize here that, in our experience, biased referees are extremely rare, much less common in fact than suspected by authors. In most cases where a conflict of interest can occur and the Editor is not in a position to know that, the referee usually declines to review the paper and explains the reason frankly. In the past several years of practice at A&A, Editors have witnessed only a few instances of obviously biased reports or delay that were the result of competition.

4.3.3   Offensive attitude of referee or author

A problem that we experience more often with referees is a slight tendency to condescension. Whether there are grounds for such an attitude or not, it is understandably upsetting to the authors; therefore, the Editors often edit reports that they perceive to be patronizing. In the few cases that border on offense, the Editors do not send the report to the author but ask the referee to re-write the report or look instead for a second, independent opinion.

Authors can also have offensive attitudes toward referees, in which case the Editors request amended replies or edit them. Again, this is a rare occurrence. A more frequent attitude on the part of the authors is to say in their reply to the report that they have taken the referee's comments into consideration, whereas the revised text is basically unchanged. This is very upsetting to the referee and can quickly lead to a stalemate in the peer-review process. To avoid this problem as much as possible, the Editors routinely request a detailed reply from the author to the referee report, including a description of the modifications to the text that were made as a result of the referee's remarks.

After this review of potential problems, we emphasize again that in most cases, the peer-review process improves articles. According to an A&A author survey organized in 2007, 28% of authors find the referee reports very useful in improving the submitted articles, 52% find them useful, and an additional 18% find them somewhat useful. These opinions justify the role of independent peer review as organized by academic journals.

4.4   The production of A&A

After a paper is accepted for publication, it undergoes a final check by the Editor-in-Chief and an English language editor who decides whether the paper should be sent out for language editing. This is an important step because A&A publishes papers originating from about sixty different countries, and English is a second language for the vast majority of authors. Homogenization of the various English idioms used by authors is thus necessary. The contribution by Joli Adams deals specifically with language editing, and I refer the reader to this chapter for details.

Once this step is over, the accepted manuscript is electronically passed to the publisher for publication. At the same time, the author is requested to send to the publisher all of the article files, since the editorial office does not deal with LaTeX files, only with PDFs[6].

---

[6]Some astronomy journals request a compilable LaTeX file and all graphical material before starting the peer-review process. The reason for this difference is mainly historical and the question of whether we should adopt the same procedure in the coming years occasionally arises. Dealing with LaTeX files at the editorial office would allow for some gain in time, *e.g.*, at the language editing stage, and the authors would also gain time at that stage since they would not implement the language corrections themselves in the LaTeX files but only okay the changes performed by language editors. However, this comes with the drawbacks that authors might lose some control over the final contents of their manuscripts and that language editors must be fluent in LaTeX,

Currently, the publisher of A&A is EDP Sciences, a small company based close to Paris that mainly publishes academic journals and books. Agnès Henri, who is in charge of the physics and astronomy department at EDP Sciences, reflects on the complex role of the publisher in scientific publishing in a subsequent chapter of this book. EDP Sciences is owned by the French Physical Society, the French Chemical Society, and one of the French mathematical societies. That our publisher is owned by learned societies insures that the company understands the needs and objectives of the scientific community. We will come back in the following to the economic model of A&A, in which the publisher plays a major role. But before doing so, we address the following question: what does the publisher do for A&A? Answering this question will hopefully show why we need a publisher.

When an author sends the publisher all the files needed to build an accepted article, they are compiled with the latest A&A LaTeX macro and thoroughly copyedited. One should perhaps clarify here the respective roles of language editors, who are part of the editorial office, and of copyeditors, who are working at the publisher. The difference between their functions is straightforward: language editors make certain that the syntax of the article is grammatically correct and consistent and that the author's writing style in English conveys the scientific message clearly and concisely, while copyeditors check the final article complies with the A&A style sheet.

The copyeditors' work often goes unnoticed by authors although it is absolutely essential for maintaining a professional quality of publication across all published articles. But who wants to deal with the fact that S/N, SNR, FWHM are in Roman characters unless they are accompanied by a specific value, in which case they must be written in italics; that there is a period at the end of a footnote (except when it ends with an internet link) but not at the end of the author's address; that Figure 3 is always abridged as Figure 3 unless it is the first word of a sentence? There are countless such rules that lead to a high level of consistency in the finished articles' style. Authors usually do not know about them, and do not have to because copyeditors will work behind the scene to make sure that all articles look the same as far as style is concerned. This thankless but essential job is done for A&A by an enthusiastic and competent team working for the publisher in Casablanca, Morocco.

After an article is processed by the copyeditors, it is sent back to EDP Sciences electronically and scheduled for publication. Several issues of the Journal, which is published weekly, are prepared in parallel to control as much as possible the balance between the different A&A scientific sections. Once an issue is complete, the editorial office does a final check of the sections in which articles appear and choose the issue cover. The issue is then sent for online publication and for printing and dispatching of the paper edition. The online publication is done by EDP Sciences, while the printing is done by a company based in Paris. On average, 35 (54) days elapse between the moment a Letter (Main Journal paper) is received by the publisher and the time it is published electronically. There is an additional 11-day delay before the printed edition is mailed to the subscribers.

---

an additional constraint for a job that is already a difficult one for which suitable candidates are rare. We are nevertheless presently testing language editing in the LaTeX files on a limited number of manuscripts to assess its feasibility in the future.

Besides publication of the electronic and printed editions of A&A, the publisher designs and maintains the A&A web site (`http://www.aanda.org`), its accompanying databases, and last but not least, the Manuscript Management System used by all actors in the peer-review system.

5 The A&A economic model

The organization of A&A is unique among physics and astronomy Journals. A&A belongs to its sponsoring countries, who finance the scientific evaluation of submitted papers through the contributions paid by the various national agencies of sponsoring countries to the Board of Directors of A&A, the entity running the Journal. Currently, the Board of Directors consists of one representative for each of the twenty-four sponsoring countries, plus one representative of ESO. Nationals from countries sponsoring A&A do not pay any additional fees or page charges for publishing in the Journal.

One should note here that the Board of Directors, although it owns A&A, is not a legal entity. ESO was therefore associated to A&A from the start and provides a legal framework for the Journal by taking care of its financial operation and holding the copyright for A&A articles. Since A&A publishes many papers from ESO researchers, the editorial process must of course remain independent of ESO to allow for an unbiased evaluation of its scientific production, meaning that the connection between ESO and A&A is thus limited to administrative and financial concerns.

The unique feature of A&A's economic model is that the operations of the editorial office, which is responsible for handling submitted articles and for improving the language of accepted articles, is financed through ESO by the national contributions of sponsoring countries and the page charges requested from authors who do not belong to A&A sponsoring countries.

The team of Scientific Editors in charge of the peer-review of A&A articles currently includes the Editor in Chief, the Editor in Chief for Letters, and eight Associate Editors located in various European countries. The editorial office staff includes two editorial assistants and three language editors. The office is hosted graciously by the Paris Observatory, who therefore also contributes to the financing of the Journal. The legal structure of the editorial office is a not-for-profit association run by members of the Paris Observatory with the help of a hired accountant, who manage the editorial office personnel and budget in accordance with French law. The editorial office yearly expenses are verified by both the ESO finance department and the ESO external auditors. Currently, the A&A Editor in Chief is also the head of the editorial office, but starting in 2012 these two functions will be separated, and the Editor in Chief will be assisted by a Managing Editor in charge of running the editorial office. The Editor in Chief will then be able to focus more on scientific and representation duties and will not need to be based in Paris.

As explained by Agnès Henri in her chapter, the Journal's web site and publication of the printed and electronic editions of A&A are done by EDP Sciences, who covers the costs of the production process with the income from subscriptions to the Journal.

A main priority of the A&A Board of Directors is to keep the subscription price (which is set in the contract negotiated with the publisher) at the lowest possible level while maintaining the highest possible quality for both the printed and online editions of the Journal. Since the subscription price is proportional to the number of published pages[7], this implies keeping this number in check. After a strong increase in published pages in the 1990s, the editorial team and the publisher went to great lengths to remain at nearly constant number of published pages yearly in spite of the increasing flow of information caused by the opening of large observational facilities.

This was done by (a) strengthening the editorial policy, (b) publishing some sections only online, (c) increasing the information content of each published page (more condensed page layout, optimized figure size), and (d) publishing some large tabular material only online. As a consequence of these efforts, the total number of published pages and the subscription price to the Journal – after taking inflation into account – has remained almost constant over the past decade, *even though the information contents published in A&A has increased by approximately 30%*. We have now reached an optimal layout that could not be changed much without altering the papers' legibility, and the number of published pages is again increasing[8] to absorb the ever-increasing information flux published in A&A.

5.1 Comparison with other journals

In contrast to other astronomy journals, A&A therefore tries to control its total number of published pages. This is seen very clearly in Figure 7, where we plot the page number evolution of the major astronomy journals. The graph does not reflect the information content of each journal, which depends on the page layout. As explained above, the information content of A&A increased considerably over the past decade but was compensated for by space-saving layout changes. Clearly, ApJ and MNRAS do not share the concern of A&A concerning the volume of publication. This is because each astronomy journal has its own economic model.

ApJ and AJ both ask authors to pay page charges and keep the subscription rate at a low level. Because the authors pay for publication, there need be no limit to the volume of publication, and the yearly page number can increase while keeping the subscription price moderate. MNRAS, on the other hand, does not request page charges from authors but has a subscription price that varies with the published volume, so the increase in page number seen in recent years is paid for chiefly by the university libraries that subscribe to the journal.

These different economic models make it difficult to compare the prices per page of the respective journals. However, an approximate calculation of expenses and revenues based on data publicly available shows that the cost per published page is roughly the same for ApJ and A&A, but the situation is different for MNRAS. In 2011, a European

---

[7]One should mention here that pages published only online are less expensive than printed pages, which explains why we are increasingly publishing only online.

[8]As a consequence, we had to publish the first issues of 2011 in the last months of 2010 in order to stay within the number of pages contracted for 2010.

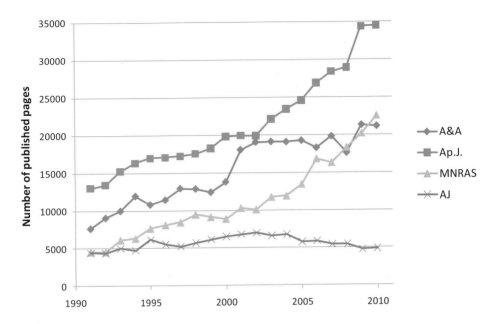

Fig. 7. Evolution of the publication volume over the last twenty years.

university library pays 5847[9] euros for the online and printed editions of MNRAS, and 5315 euros for the printed edition only. Surprisingly, MNRAS does not offer the choice of subscribing only to the electronic edition. In comparison, the subscription prices of A&A are 4015 euros for the printed and electronic editions and 2915 euros for the electronic-only edition. Consulting MNRAS online is therefore more expensive by a factor of 2 than reading A&A online, although both journals publish a similar number of pages each year. The comparison is not entirely fair, however, since one should take the national contributions requested from A&A sponsoring countries into account. Doing so, we find that the contribution adds approximately 500 euros to each subscription paid by the main sponsors of A&A. Even when taking this additional, somewhat hidden cost into account, the price of the A&A subscription remains substantially lower than that of MNRAS. Whether the continuous price increase in the MNRAS subscription price is sustainable without losing institutional subscribers remains to be seen.

5.2   Full open access for A&A?

The term "open access" has been a buzzword in the scientific publishing world for a number of years. What is it exactly? As the online encyclopedia Wikipedia[10] notes, the concept comes in a variety of flavors:

---

[9] Prices indicated here do not include the VAT.
[10] See http://en.wikipedia.org/wiki/Open_access_(publishing).

"Open access (OA) refers to unrestricted online access to articles published in scholarly journals, and increasingly also book chapters or monographs. [...] OA can be provided in two ways:

- "Green OA" is provided by authors publishing in any journal and then self-archiving their postprints in their institutional repository or on some other OA website. Green OA journal publishers[6] endorse immediate OA self-archiving by their authors.

- "Gold OA" is provided by authors publishing in an open access journal that provides immediate OA to all of its articles on the publisher's website. (*Hybrid open access journals* provide Gold OA only for those individual articles for which their authors (or their author's institution or funder) pay an OA publishing fee")[11].

Where do we stand at A&A concerning these various degrees of OA? Currently, an issue of A&A is in Gold OA for the week following its publication. This allows astronomers who do not have a subscription to A&A to read and download the most recent publications in their areas of scientific interest. Our publisher monitors the downloads to make certain that there is no foul play by some institutions. Once this period of Gold OA is finished, A&A becomes a hybrid open access journal, since authors can pay a nominal fee to have their article in OA; the A&A Letters are in Gold OA at all times, as are the online sections (astronomical instrumentation, online data and catalogs, numerical methods and codes, and atomic and molecular data). The reason behind this policy is to provide Gold OA for those sections that are of interest to a relatively wide community. Furthermore, A&A is a *delayed open access journal* in the sense that, after two years, all published articles can be read by non-subscribers and are thus again in Gold OA. Furthermore, our publisher encourages ArXiv archiving or self-archiving of articles published in the Journal without any period of embargo, so that A&A is at all times a Green OA journal. As a consequence of the generous OA policy of EDP Sciences, A&A is currently the most open journal in astronomy. Luckily there has been no significant drop in subscriptions since the policy was put in place. If the number of library subscriptions declined because of it, our publisher would have to come back to a less open policy.

Why can't A&A provide Gold OA at all times? The reason is obvious: publishing the Journal has a cost that someone has to support. As we explained above, the sponsoring countries and page charges levied from non-sponsoring authors pay for the peer-review process, editorial office costs, and language editing. To provide full open access to A&A, one must find a way to cover the costs of the publisher, *i.e.*, the copy-editing of articles and production of the online edition of the Journal including web sites and archival.

We could provide full open access to A&A online to all sponsoring countries if we were to levy page charges on everyone, A&A sponsor or not, so that the page charges cover the production costs of the online edition. A scheme including a reduction of page charges for authors from sponsoring countries could be worked out to encourage sponsorship of the Journal. Is it a better scheme than the current subscription-based financing? It seems sensible that the funding agencies who pay billions of euros in salaries and in new instrument developments every year be asked to cover for the comparatively very small

---

[11]Released under license; *cf.* http://creativecommons.org/licenses/by-sa/3.0/

cost of publication of their research results through the page charges. Because the printed edition of A&A would not exist anymore (except for archival purposes), the production cost of the Journal would be lower and the free access for all would allow for the widest possible dissemination of the research results published in A&A. Working toward open access for a fully online journal thus appears to be a worthwhile goal for the near future.

6 Conclusions

We have discussed various aspects of *Astronomy & Astrophysics*, one of the leading journals in astronomical research, as an example of what peer-reviewed publication consists of in our area of scientific work. Although they may differ in their details, the operations of the other astronomy journals are very similar to those of A&A. In particular, the acceptance criteria for published articles are very much the same for the four main journals, a homogeneity that results directly from the small size of our community. As a consequence, the impact of these four journals, which publish more than 95% of the global astronomical production, are similar, with impact factors ranging from four to six. This contrasts strongly with some other scientific fields, such as molecular biology, where impact factors of the various journals span a range of less than one to more than fifty.

One question that surfaces again and again in our community is whether we still need journals. Why can't we simply put our results on astro-ph, and let the readers decide what is sensible and what is dross? One argument in favor of journals is trivial: who has time to read badly argued papers for extracting the useful information that might lurk in them? Journals act partly as a filter for the publication flow that recycle poorly organized material into better structured, useful pieces of work. The second argument is complementary: as shown above, peer review does improve the papers. Authors acknowledge the useful role of referees, although they might be tempted to vilify them at times.

The validating role of the research results played by academic journals is in fact so important in the scientific process that it should logically lead to recognizing that journals are an integral part of the research infrastructure, on the same level as scientific instruments and archival databases, which would justify financing of community-owned journals by the science funding agencies. This in turn pleads for the idea that journals are too important to be left in the hands of commercial publishers, whose profit-oriented approach is hardly compatible with the values of the scientific community. The open access initiative was in fact prompted by the excessive subscription prices to academic journals imposed by some big publishing houses. This does not mean, however, that we should get rid of publishers altogether. We have seen that their role is essential in shaping professional-looking journals that do justice to the exciting research results published in their pages. However, the journals ought to associate with low-profit publishers (*e.g.*, university presses, community-owned publishing houses) and work with them toward establishing fair prices allowing for a reasonable cost of publication and wide, free dissemination of the scientific results.

References

Bertout, C., & Schneider, P., 1994, A&A, 420, E1

Pottasch, S.R., 1999, `www.aanda.org/content/view/103/107/lang,en/`

Scientific Writing for Young Astronomers
C. Sterken (ed)
EAS Publications Series, **49** *(2011) 23–31*
www.eas-journal.org

THE HISTORY OF THE CREATION OF *ASTRONOMY & ASTROPHYSICS*

S.R. Pottasch[1]

Abstract. Dutch astronomer S.R. Pottasch, who was appointed as one of the two first Editors-in-Chief of A&A together with J.L. Steinberg, recalls the events that led to the creation of the first European astronomy journal. This account is based on the author's notes and documents and also on an article relating ESO's early history written by A. Blaauw and published in *The ESO Messenger* No. 60, pp. 23–34 (1990). C. Bertout added footnotes giving a few biographical details about the actors.

History began with the meeting of 8 April 1968 at the Leiden Observatory. Everything that happened before that date can be considered as prehistory. It is essential to understand the prehistory to make sense of the history, but it is not easy to describe. This is because it consists of unrecorded discussions of individuals in small groups, feelings concerning whether one's papers are being read, etc. And all this must be put in the context of a rapidly expanding European astronomical community with an outdated publication structure. In the early 1960's European publication was splintered. In the best cases each country had its own publication(s); this was sometimes also true of smaller countries. In addition most observatories had their own publications, which were sometimes simply reprinted articles that appeared elsewhere, but sometimes were the only source of original work. This system worked more or less when there was a limited number of astronomers; in that case it was possible for authors to know who was working in their particular field, and to send reprints to these persons. As the number of astronomers increased, this system began to break down. It is easy to imagine that a substantial number of astronomers blamed the publishing situation for the fact that their work was not having the impact that they expected. The following journals were active in western Europe at that time (with the date of founding in parenthesis)

- Annales d'Astrophysique (Ann. d'Ap., 1938) published in French with occasional English;

- Bulletin Astronomique (1884) published in French;

---

[1] Kapteyn Astronomical Institute, PO Box 800, 9700 AV Groningen, The Netherlands

DOI: 10.1051/eas/1149002

- Bulletin of the Astronomical Institutes of the Netherlands (BAN, 1921) published in English;

- Journal des Observateurs (1915) published in French;

- Monthly Notices of the Royal Astronomical Society (MNRAS, 1895) published in English;

- Zeitschrift fur Astrophysik (ZfAp, 1930) published in German with some English.

In addition other journals existed in Italy, East Germany, Scandinavia, Poland and Czechoslovakia. The number of subscriptions to these journals was roughly 300 to 450 except for Monthly Notices, which was higher. This probably was due to the large number of amateurs who subscribed and not to a substantially larger number of astronomers. The Astrophysical Journal (ApJ), on the other hand, did have substantially more astronomer subscribers. While many of these subscriptions were American, personal subscriptions to the ApJ by European astronomers were much more common than personal subscriptions to European journals by Americans.

The financial situation of each of the European journals was different. The French journals were completely financed by the government (CNRS). ZfAP was owned and financed by a private publishing house (Springer Verlag). The Dutch journal was financed from the budget of the individual Dutch observatories, while MNRAS was owned and financed by the Royal Astronomical Society. The choice of the editor and the editorial policy were thus not always in the hands of astronomers.

When I came to Groningen University in 1963, I was appointed Editor-in-Chief of BAN by the directors of the Dutch astronomical institutes. In the course of the first year in this function, I became aware of several important shortcomings in the journal. First of all, there was not an adequate refereeing system. The articles were submitted through the directors of the astronomical institutes who were responsible for the scientific content. This is a form of refereeing that works better in small institutes than in large ones. Secondly, the publication time was sometimes long. This was because it was necessary to wait with publication until enough articles were submitted to fill an issue. This could cause a delay of six months in publication, but it was unpredictable. Thirdly, because the Dutch astronomical community was small and specialized, the BAN was read more often only by astronomers in these fields. This led to a growing feeling that the journal was not widely read and that when one had an important result it should be published elsewhere. A final difficulty was that the journal was distributed free of charge to the staff and students. Because of the easy access the students consulted the BAN too often, and the other astronomical journals too little. It generated "inbreeding".

Jean-Louis Steinberg became Editor-in-Chief of Ann. d'Ap. at about the same time (1963). We had known each other quite well since 1959 when I was working at the Meudon Observatory. We met several times in the course of the following years to discuss the problems of publication. Ann. d'Ap. had problems similar to those faced by the BAN with the additional problem that the widespread use of the French language discouraged certain groups of astronomers from reading it. We decided to investigate the possibility of combining the two journals, and initiated discussions in our respective countries. As might be expected, the initial reactions were reserved, but still encouraging. One of the reservations was whether the combination of only the two journals was a big enough step considering the difficulties that were involved. One encouraging reaction was from the director of publications of the CNRS, Y. Mazière, who saw the possibilities in internationalizing French publications and who lent his full support to the project.

At about the same time I received an invitation from A. Reiz, at that time the director of the Copenhagen Observatory, to visit with the purpose of discussing the possibility of combining the Scandinavian journals and Observatory publications with the BAN. The spontaneous interest of the Scandinavian astronomers led to contact with the Swedish ministry, whose representative, Dr. G.W. Funke, immediately took an active interest in the formation of a European journal. A meeting took place in February 1967 in Copenhagen. Present were the directors of all Scandinavian astronomical institutes (4 Swedes, 3 Danes, 2 Finns, Jensen[1] representing Oslo, and Funke) and myself, representing the Dutch. There was agreement in principle to establish a joint Scandinavian-Dutch journal, and a meeting to work out the details was to take place later in the Netherlands. This meeting never took place because the possibility of establishing a larger journal took precedence. The Scandinavians made it clear, however, that if a larger journal was not set up, they would establish their own journal.

The two major countries for which problems existed were Germany and Great Britain. The situation in these countries was similar in the sense that both had existing astronomical journals. There was an important practical difference, however, since ZfAP was owned by its publisher, while MNRAS was owned by the Royal Astronomical Society. This difference was, at least to some extent, responsible for the way in which the German and British astronomers viewed the situation. The German astronomers who were most active in this matter are the chairman of the Council of West German Observatories, L. Biermann[2]; the Director of ESO and of the Hamburg Observatory, O. Heckmann[3]; and W. Fricke[4]. All were strong proponents of a European journal. They all felt that it was not a strong priority to try to incorporate ZfAp into the European journal. They felt that the German ministry would financially support the new journal and that the ZfAp would eventually disappear.

The British astronomers looked at the situation differently. The most important contact was D.H. Sadler[5], director of the Greenwich Observatories, who was very sympathetic to the idea of a European journal. Another was F. Graham-Smith[6]. The subject was discussed at a meeting of the Council of the RAS on 13 October 1967, after arguments were assembled by letter. A merger of MNRAS with other journals was rejected. The arguments given were (1) that MNRAS was an old journal with a long tradition, and (2) that the Royal Astronomical Society would have little function if MNRAS were given

---

[1] Eberhart Jensen (1922–2003), then director of the Oslo Solar Observatory.

[2] Ludwig Biermann (1907–1986) was the director of the Munich Astrophysics Institute (now the Max-Planck-Institute for Astrophysics).

[3] Otto Heckmann (1901–1983) became the first ESO Director General in 1962.

[4] Walter Fricke (1915–1988) was the director of the Astronomisches Recheninstitut in Heidelberg from 1955 to 1985.

[5] Donald Harry Sadler (1908–1987).

[6] Sir Francis Graham-Smith, born in 1923, was professor of radioastronomy at Manchester.

up. This last argument probably weighed heavily, since selling MNRAS was by far the largest source of income for the RAS. This decision was painful because MNRAS was highly regarded by the continental European astronomers.

This had been the second painful event that had occurred in 1967. The first was the announcement by Z. Kopal[7] (in a letter dated June) that a new journal had been formed. Besides Kopal, very few astronomers were involved in its formation. The financial basis for this journal was the willingness (perhaps desire) of the Reidel Publishing Co. to do the publishing. Reidel was at the same time owner of the journal. The new journal, called "Astrophysics and Space Science" was not the only astronomical journal published by Reidel, and there was some suspicion that it was formed on Reidel's initiative. Though this has been denied, the project was kept completely secret until Kopal's announcement, for what Reidel called "commercial reasons". It was hard for astronomers to see in this new journal a replacement for the existing national journals for several reasons. First of all, it was seen as a commercial undertaking in which the wishes and desires of the astronomers were subordinate to those of the publishers. Furthermore, there were no checks on the scientific quality, since the editor was appointed for an indefinite term by the publisher. For these reasons various astronomers (among them J.H. Oort[8] and J.C. Pecker[9]) wrote letters to Reidel and to Kopal protesting the formation of the new journal, and using as argument the ongoing discussions on the merger of the European journals. Kopal's answer was interesting. In a letter dated 7 August 1967 addressed to Oort, Kopal says "As to the possibility of an European journal, I heard the idea being mooted about at least for 10–15 years now – usually to forestall some other budding development; and when this aim has been accomplished the project went back again to slumber. I am afraid Great Britain will sooner join the European Common Market than all west-European astronomers will agree on a joint journal; neither you nor I may alas live long enough to see it happen".

Contrary to Kopal's insight, the European astronomers were serious. At the beginning of November 1967 the Dutch astronomers met to discuss the situation. They concluded that it would be desirable to form a new journal sponsored by France, the Netherlands, Belgium, and Scandinavia. They added the stipulation that the European Southern

Observatory be asked to also sponsor the journal. They agreed to give up the BAN when the new journal was formed. I was asked to inform the French of this, which I did through contact with Steinberg. He had already prepared a note on the subject, in which he came to the conclusion that a merger of European journals was not only desirable but necessary. With this note as a basis, Steinberg and Denisse[10], called a meeting of the French astronomers. The meeting, which took place in December 1967, was attended by all astronomers with a Ph.D. level or higher. The result was that 75 present agreed that a new

---

[7]Zdenek Kopal (1914–1993), was the head of the astronomy department at the University of Manchester. He remained Editor in Chief of Astrophysics & Space Science from 1968 to his death in 1993.

[8]The highly influential Dutch astronomer Jan Hendrik Oort (1900–1992) was professor at the Leiden Observatory.

[9]Jean-Claude Pecker, born in 1923, was the director of the Nice Observatory.

[10]Jean-François Denisse, born in 1915, was director of the Paris Observatory from 1963 to 1967.

journal was desirable. Steinberg and I then prepared a proposal that can best be described as the "ground rules" for a new journal. This proposal was then sent to Reiz, the chairman of the Scandinavian group for their opinion. Steinberg contacted the Belgians. In a less formal way, several German and Italian astronomers were also informed.

At this point it was necessary to discuss the matter in a more formal manner. In a letter dated 27 March 1968 I invited a number of astronomers to a meeting on 8 April 1968 at the Observatory in Leiden. The purpose of the meeting (as stated in the letter) was fourfold:

- to discuss the general principles under which the new journal would operate;

- to discuss the financial implications of the new journal;

- to begin a discussion of the detailed operation of the new journal;

- to agree on a date for beginning the new journal.

Those who attended the meeting, aside from Steinberg and myself, were Reiz (representing Scandinavian countries), van Bueren[11] and Oort (The Netherlands), Ledoux[12] (Belgium), Schatzman[13] and Mazière (France), and Biermann (Germany). Velghe[14] and Swings[15] (Belgium), and Funke (Sweden) wrote that they were sympathetic with the plans but were unable to come. In addition Funke wrote that the Swedish Research Council was interested in sponsoring the journal. ESO representatives Heckmann and Blaauw[16] were not in Europe, but indicated that ESO would like to co-sponsor the journal. The minutes of the meeting were taken by Mrs. Ondei, Oort's secretary. All persons present saw the desirability of merging the existing journals to form a new journal, on the widest possible European basis. It was proposed that the new journal be called "Astronomy & Astrophysics, A European Journal". Steinberg stated that the French astronomers had agreed to stop publication of Ann.d'Ap when the new journal began publication, but the CNRS still had to approve this step. Oort stated that the Dutch had in principle agreed to stop publication of the BAN.

The main policy-making body of the new journal was to be the "Board of Directors", consisting of senior astronomers or government representatives of the sponsoring countries. It would normally meet once a year to discuss the yearly report of the Editor and the state of the journal. It would also elect the Editor(s) and approve the appointments to the Editorial Board. Each country was responsible for the appointment of its member(s). New sponsoring countries would mean an expansion of the "Board of Directors". Most of those present felt a single editor, appointed for 5 years, was optimal. However in the initial stages of the journal two editors of different nationalities were proposed, which was to ensure that no individual country would dominate the journal.

---

[11] H.G. van Bueren was the director of the Utrecht Observatory.

[12] Paul Ledoux (1914–1988) was professor at the Liège Astrophysical Institute.

[13] Evry Schatzman (1920–2010) was at that time the head of the Meudon Astrophysical Observatory.

[14] Albert G. Velghe (1916–1986) was then director of the Royal Observatory in Brussels.

[15] Pol Swings (1906–1983) was professor at the Liège University.

[16] Adriaan Blaauw (1914–2010), director of the Kapteyn Institute in Groningen, became the second ESO Director-General in 1970.

Articles could be submitted by citizens of any country in the world. Acceptance would depend only on its scientific quality. There would be no page charges. The journal would be the property of the astronomers, represented by the "Board of Directors", who would be responsible for the scientific content. A publisher would be contracted to print and distribute the journal. He would be responsible for the technical quality of the journal and for its promotion and sales.

Because the "Board of Directors" had no legal status, a number of problems arose. For example, the CNRS was not able to send funds to a private organization outside of France. And who had the authority to sign a contract with the publisher? To circumvent the possibility that a treaty among the sponsoring countries would be necessary, which would involve several years delay, it was thought desirable to involve ESO as one of the sponsors, and let ESO handle the finances and the legal side of the journal. The choice fell on ESO, although the IAU and ESRO (the present ESA) were also considered, both because most of the countries acting as sponsors were also members of ESO and vice-versa, and because ESO was sympathetic to the proposed journal. At the time of the meeting, only informal contact had been made with ESO, and although the initial response was positive, the matter was to be formally discussed by the ESO council at its meeting in July 1968.

The question of who should publish the journal was raised. Two offers had already been made, one from Reidel Publishing Co., the other from North Holland Publishing Co. The one from Reidel was 10–15% cheaper, but after much discussion it was decided not to consider Reidel further as long as it continued to publish Astrophysics and Space Science. Furthermore their ability to publish the journal for a longer period at the price quoted was questioned.

It was agreed that the following "countries" would support the journal financially and would be represented on the Board of Directors: France (with 4 representatives), Netherlands (2), Scandinavia (2), Belgium (1) and ESO (1). Concerning Germany, the situation was less clear. Biermann stated that although ZfAp would continue for the time being, it did not preclude German participation as a sponsor of A&A. He further indicated that many German astronomers, especially of the younger generation, favored an international journal. All present felt that it would be important to have German participation right from the beginning. Steinberg and I (as editors of the two largest journals which would cease publication) were asked to write a letter to the Council of West German Observatories requesting it to sponsor the new journal. Germany would be free to decide on the number of representative it would have on the Board of Directors.

It was considered desirable to begin publication of the first issue of A&A on January 1969. The first volume was to be in the hands of the publisher by 30 September 1968 at the latest, and was to be refereed before that date. The future timetable was agreed upon: an agreement by the (astronomical) councils who acted as sponsors and financiers was to be obtained by the end of June. The Board of Directors was to meet directly after the ESO meeting to appoint the Editor(s), to decide on the publisher and consider the appointment of an Editorial Board. The journal was beginning to take shape, and quickly.

The next meeting took place on 3 July 1968 in the University Foundation, Egmontstraat, Brussels. This was scheduled to coincide with an ESO Council meeting in Brussels, which had just been completed. Although decisions were to be made at this meeting, as they were at the April meeting, this is not yet considered as the first meeting of the Board of Directors. The participants were the same as at the Leiden meeting, except for van Bueren and Ledoux. In addition, Heckmann and Blaauw (ESO), Velghe (B), Fehrenbach[17] and Kovalevski[18] (France) and Voigt[19] (Germany) were present. The minutes were taken by Miss Geier (ESO) and I was chairman.

The situation at the beginning of the meeting was as follows.

- France: the CNRS had endorsed the merging of all major French astronomical publications in A&A, and has agreed to pay a $ 24,000 subsidy. The French delegates to the Board were to be Mazière, Delhaye[20], Schatzman and Fehrenbach.

- The Netherlands: The Dutch Astronomical Council had agreed to merging the BAN in A&A, and has agreed to pay a $ 10,000 subsidy. The Dutch delegates were Oort and van Bueren.

- Belgium: The Belgian delegate was Velghe, but it was not yet official. The subsidy was to be $ 3,000 in 1969.

- Scandinavia: The delegates were to be Strömgren[21] and Funke, and the subsidy $ 10,000.

- Germany: The Germans had now agreed to sponsor A&A. Biermann, Fricke, and Unsöld[22] were to be the delegates to the Board. The method of paying the subsidy was still under discussion. The possibility of merging ZfAp depended on negotiations underway with Springer Verlag.

- ESO: The ESO Council agreed to make its administrative and legal services available to A&A. The ESO Council had authorized its Director to conclude (1) a contract with A&A, and (2) a contract with the publisher. It would therefore be possible for ESO to act as financial agent for the Board, although the financial responsibility remained with the Board. ESO would not contribute financially to the journal, but did not charge for "legal" services rendered to A&A.

The situation as regards to a publisher had changed somewhat since the last meeting. There were now four publishing houses that had made offers: Springer Verlag, Reidel, North Holland, and Wolters Noordhoff. Reidel and Wolters Noordhoff were the cheapest, but the reason for not considering Reidel further had been discussed in the previous meeting. The Board considered Wolters Noordhoff too inexperienced. Springer had indicated in its bid that if it were chosen it would stop publication of ZfAp. Since this would

---

[17]Charles Fehrenbach (1914–2008) was the director of the Haute-Provence Observatory.

[18]Jean Kovalevsky, born in 1929, astrometrist at Nice Observatory.

[19]Hans-Heinrich Voigt, born in 1921, was then the director of the Göttingen Observatory. He later became A&A Editor in Chief together with J.-L. Steinberg and S.R. Pottasch (Letters Editor).

[20]Jean Delhaye (1921–2001) was the director of the Paris Observatory from 1967 to 1971.

[21]Bengt Strömgren (1908–1987) was an influential Danish astronomer who spent a large part of his career in the USA (Mc Donald Observatory and Princeton).

[22]Albrecht Unsöld (1905–1995) was the director of the Institute for Theoretical Astrophysics at Kiel University and Editor in Chief of ZfAp.

enhance the chances of success for the new journal, it was attractive. But the meeting felt that this was not a strong enough reason to accept a higher price. Heckmann, voicing the sentiment of the meeting, proposed contacting Springer again to negotiate a price reduction. He stated that "One should not be afraid of dropping Springer, as ZfAp will slowly die once the new journal is formed". The meeting was adjourned for 20 minutes, while a telephone call was made to Springer. This resulted in the agreement by Springer to lower its bid to that of the lowest bidder. Detailed negotiations were to take place in the near future. Although it was not until October that a final agreement was reached, Springer was hereafter considered as the foremost candidate for publisher of the journal.

At this point another problem arose. Kovalevski, speaking for a group of fundamental astronomers mostly from France, proposed to split the not yet formed journal into two parts: a large one for astronomy and astrophysics, and a smaller one for celestial mechanics. The reason given was the specialized needs for publication in this field. While the meeting was sympathetic to this view, soon after Fricke objected to such a division, since it might lead to a subsequent splitting of the journal into a number of over-specialised journals. Eventually Fricke, Kovalevski, and Morando[23] prepared a short paper in which their conditions for publishing on Fundamental Astronomy and Celestial Mechanics were stated. This subject became one of the original sections of the journal, and no further problems were forthcoming.

Toward the end of this meeting the first Editors-in-Chief were appointed. Oort took the initiative in proposing to the "Board of Directors" that Steinberg and I should take over this task. We accepted, and later prepared a memorandum on the division of the tasks between us. With this step the date of 1 January 1969 for the start of the journal became more realistic. The biggest problem now was to finalize the contract with the publishers.

The following meeting is officially regarded as the first meeting of the Board of Directors. It took place in Paris on 11 October 1968. The meeting began with a word of welcome by the Director General of the CNRS, Professor Jacquinot. The meeting had been called on short notice because Mayer-Kaupp, who was in charge of the negotiations for Springer had been away until 23 September, and it was only useful to hold the meeting if a reasonable draft contract was available. Because of the short notice, five of the eleven Board members were unable to attend, so that it was decided that all decisions made were to be provisional.

The first act of the Board was to elect A. Blaauw as its first Chairman. Mazière was elected as Treasurer of the Board. The main business of the meeting was to discuss the draft contract with the publisher. The advice of the ESO lawyer, Walters, was also available. The Board went through the contract, article by article, and made changes where it saw necessary. Then Mayer-Kaupp, who was present in the building, was invited to attend the meeting and given the proposed changes. He accepted them, and the new contract was then adopted by all the Board members present. The contract, between ESO and Springer, would be signed at the next meeting. Since it was now already October and the first issue of the journal was scheduled to appear in January, this had to be a formality.

---

[23] Bruno Morando (1931–1995) was the director of *Bureau des Longitudes* in Paris.

Other matters discussed were (1) an agreement between the Board of Directors and the Editors, (2) a draft agreement between the Board and ESO, (3) the matter of Sub-editors (it was decided that there would be no Sub-editors), (4) the composition of the Board of Editors, and (5) the sections which would be listed in the table of contents. Finally the matter of the Supplement Series was discussed. This publication was to be independent of the Main Journal and Springer, and was to be published by the Board of Directors. It was decided to approach L.L.E. Braes[24] of the Leiden Observatory to ask him to take charge of the publishing. All articles would be submitted to the Editors-in-Chief and would only be sent to Braes after they had been refereed and accepted. It was not clear at this time where the binding and mailing would take place. In contrast to the Main Journal, a page charge was to be levied. The first issue of the Supplement Series appeared in January 1970.

It was clear that a new meeting would be necessary within the near future, if only to sign a contract with the publisher and make certain that no insurmountable problems made it impossible for A&A to appear on time. Therefore the 2nd meeting of the Board of Directors took place on 2 December 1968 at the ESO headquarters on the Bergedorfer Strasse in Hamburg. An interesting sidelight of this meeting was the interest shown by two other countries to become part of the journal. Switzerland sent the director of the Geneva Observatory, M. Golay[25], as observer, and Italy sent the director of the Trieste Observatory, M. Hack[26].

This meeting gave an official form to the journal. A contract was signed between the Board and ESO, which now gave a legal status to the decisions of the Board, and enabled it to take financial responsibilities. A contract was signed between ESO (acting for the Board) and Springer Verlag to publish the journal for a five-year period. A contract was signed between the Board and the Editors for a three-year period. A memorandum was prepared by the Editors on the sharing of the responsibility for the editorial work, which was discussed by the Board. The Standing Rules for the Board of Directors were approved, which included rules for membership expansion, and financial contributions. B. Strömgren was elected vice-chairman. All was now complete: the first issue of Astronomy & Astrophysics was to appear 5 weeks later.

Groningen, October 1999

---

[24]Luc L. Braes, born 1936.
[25]Marcel Golay, born 1927.
[26]Margaretha Hack, born 1922.

Scientific Writing for Young Astronomers
C. Sterken (ed)
EAS Publications Series, **49** *(2011) 33–61*
www.eas-journal.org

THE EDITOR-REFEREE SYSTEM AND PUBLICATION AN EDITOR'S VIEW OF THE PROCESS

S.N. Shore [1]

Abstract. This chapter explains the functioning of scientific journals from the editorial side of the process. Both the history and current functioning of scientific journals are reviewed with a particular emphasis on the evolution of the referee's role. In its current form, the evaluation of a submission is interactive between the three parties – the author(s), editors, and reviewers. The editors serve as the mediators and final evaluators, seeking advice from one or more contacted experts who are in the special position of evaluating the science, presentation, and significance of the work. The chapter explains how this proceeds, and its advantages, pitfalls, and criteria – scientific, archival, and ethical – and how these have evolved historically and consensually. Since referees and editors are also authors, the symbiosis of the process is one of its strengths, since all participants exchange roles.

1 Introduction

The editor plays two primary roles, to choose an appropriate reviewer for a paper and to insure that the process is fair, civil, and in the best interests of both the journal and the parties concerned. Above this there is another function, perhaps less well understood, that of actually accepting the paper for publication. This chapter, written from the point of view of the editors, is meant to serve several purposes: to explain how the editorial process functions in scientific publication, some of its history, and the role of the referees. It is very likely that you will eventually be called on to step to the other side of the publication process, to evaluate your peers' work, so before that happens, it is important to step back and consider why the process exists, how it functions now, and how it changes.

2 Evolution of the editorial-referee system

2.1 The origin of scientific journals

The history of scientific communication may seem outside the purview of this chapter but it is relevant to understanding how the different features of scientific publication have

---

[1] Scientific Editor, *Astronomy and Astrophysics,* Dipartimento di Fisica "Enrico Fermi", Università di Pisa, Italy

evolved. And because science as a discipline began in the universities, which were themselves founded in the Middle Ages, I start there. Originally publication meant depositing the original manuscript of lectures (an *exemplar*) with a scribal studio licensed by the university (and therefore approved by both the faculty and hierarchy) who had the right to make, distribute, and, and charge for, copies on request. These could be any portion of the manuscript or the whole work, and the material was known to be intellectual product and the rights – that is, the assignment of authorship – was implicit in the consignment of such material. The authors, who were already being paid for the lectures and examinations, generally received nothing from the copyists, nor did they have to pay for the service. It was understood that the rights (such as they were) belonged to the copyist.

This changed with the introduction of printing in the 15th century. Publishers went to considerable expense to create their houses, to hire trained staff, and now also to procure the necessary equipment (*i.e.*, type, the press itself, paper) and assumed the right to sell the works they produced. But to continue their business, they were required to be officially sanctioned, and the individual works had to be submitted to judgment (the *imprimatur* of the local higher clerics) since the work could now be distributed in a more complete form and was not just for personal use. A book, once published, had a far wider circulation with many more copies than a manuscript and each copy was precisely the same. There were attempts at censorship: the *Index* – an ecclesiastic list of forbidden reading within theological doctrine – was created in the mid 16th century to inhibit circulation where publication could not be prevented (generally because the publishing houses were outside of the Papal states, especially in Holland and Venice).[1] However spectacular the individual cases of attempted suppression, the literature was unstoppable.

The intellectual and publishing landscape changed in the middle of the 17th century with the founding of philosophical, that is scientific, societies. These received not just sanction but also patronage. They controlled the membership, being self-governing, which was the forum of communication. The proceedings of the meetings were recorded, in various forms and to enhance the connectivity of the different learned societies, these records were exchanged. Thus were born the *journals*, the very names of which preserve the nature of the enterprise, *e.g. Comptes Rendus, Philosophical Transactions*, various *Acta*, and so on. The society retained publishers, or assigned the task of publication to individual members (the secretaries), who undertook the production of the reports as the first edited proceedings. These could be quite elaborate, detailing both the discussions and the presentations, and often appeared in annual collections or even extending over several years because of the slow pace of production. Some publishers developed particular specialties, such as an expertise in setting mathematics (with problems similar to Greek or Hebrew texts that required special treatment and types, or music) and often these technical works were published by the same houses as the other humanist productions. The editor overseeing the production often took an active role in production that

---

[1] This did not mean the books were necessarily suppressed, often particularly offensive sections were blacked out as, for instance, in *de Revolutionibus* of Copernicus. Those portions not contradicting doctrine were permitted reading, even to the extent of allowing the material to be taught in university lectures. The pursuit of "theological offenders" became more furious in the late 16th century, Giordano Bruno being the principal example in philosophy, and continued through Galileo's trial before the Inquisition.

went beyond verifying the accuracy of the text, one that included communication between the publisher and the authors. This is especially famous in the case of the three editions of Newton's *Principia*, ordered printed by the Royal Society of London and overseen by Edmond Halley. In this role, Halley established himself not only as Newton's drudge but also as an amanuensis, an aid who actually refereed the material, querying Newton at various, often critical, points in the text and requesting extensions or clarifications, or pointing out errors. Robert Cotes served this role in the second edition with even more attention to the details. The editors of the proceedings were, in general, the founders or principals of the societies and as such were in the best position to judge the contents when deciding on publication. But members of the society had the right, by membership, to have their material appear in the proceedings and also in separate works sanctioned by the society. These society proceedings continued through the 18th century to be the main source of current discussion, playing much the same role as the bulletins of meetings – the abstract books – play now. They provided a snapshot of the state of the field at the time. As the membership grew and the finances became more secure, the works appeared with greater regularity and frequency. By the end of the century, there were several dozen in circulation along with an increasing monographic output.

One singularity stands out in the 18th century, the Académie Royale des Sciences (Paris). Unlike the English Royal Society, the academiciens established a separate publications committee, the Comité de Librairie, that served as the adjudicating body for all papers published in the *Histoire* and *Memoires* (McLelland 2003). Even communications to the academy were subjected to critical review before being consigned to the printer and, more important, those outside the academy (called foreigners, not in the sense of nationality but membership) were permitted to submit papers. The significant innovation of the Comité was to create of a set of rules – standards for review and acceptance – that read almost precisely like those later recreated by the modern journals in the 20th century. For instance, a member had to assure the Comité that the paper had not been either submitted or published elsewhere, and papers were refused if they were not sufficiently novel or different from those already published. Expert reviewers were appointed by the Comité to evaluate papers of special interest or in areas in which the members felt less able to judge expertly. This often applied to papers in mathematics and, especially, astronomy. It was not enough that a result be novel, however, it had to be presented in a consistent, complete way that permitted its evaluation. Priority disputes, plagiarism, and proper citation were all considered in the criteria, any of which could suffice to reject a paper even if it had been read before to the Académie. This, unlike the Royal Society, presentation at a meeting did not insure the eventual appearance of the complete work but it was a minimum criterion (that is, public presentation).

2.2   Enter the referees

This changed in the 19th century in several ways, which are fundamental to our world. First, new institutions and societies were founded, more specialized in their nature and more democratic in their membership that did not require particular credentials other than the scientific activity itself and were not class-based, as was the Royal Society. The Manchester Lunar Society, for instance (so named because it held monthly meetings) included practical engineers (*e.g.* Watt, Wedgewood), and the Royal Institution (founded mainly for public outreach, as we'd now put it) opened scientific discussion to the literate masses who began to take a more serious interest in science. This new audience also desired outlets for their interests, and besides more general forums such as the Proceedings wanted their own, less "controlled" (for that read class-laden) outlet. This market was supplied, very soon, with private journals, such as A.L. Crelle's *Journal für die reine und angewandte Mathematik* (known almost universally as *Crelle's Journal*) in Germany and the *Philosophical Magazine* in England. These were under the control of the editors themselves, not any society, and since their funds were supplied (for profit) by subscription they needed to have the highest quality, most interesting material possible.[2] Even among the oldest, best-established publications, there was a growing interest in reaching new readers, by extending the base.

The editors of the *Phil. Trans.* and *Phil. Mag.* were among the best in their field but, given the diverse nature of the material and the ceding of content control to the editors by the various organizations, they found it necessary to get additional advice on the submissions. This was the beginning of the formal refereeing system. The Royal Society, that had maintained a similar body to that of the Académie, the "Committee of Papers", changed to independent referees in 1831 whose function was to independently review the suitability of papers for publication and formally submit a report to the Committee (Anderson 1993). For the *Phil. Trans.*, for example, there are extensive records of the referee reports. For instance, Jackson & Launder (2007) discuss the reports on various papers by Osborne Reynolds (he of the Reynolds number) for which the editor, G.G. Stokes (he of the theorem), maintained a close circle of advisors who could judge papers and, in anonymous reports, recommend acceptance or rejection of the submission including Maxwell, Rayleigh, and Kelvin. The final decision, ultimately, rested with the editor but the reports were the fundamental element in the decision. While this also included the judgment of the editor himself (alas, the gender was unique at the time), it was more usually one of the reviewers. The journals also served as proxy meeting sites, publishing commentaries, rebuttals, exchanges in public that would normally have been accessible only in meetings.[3]

The next innovation was the weekly and monthly scientific *magazines*, again purely commercial ventures undertaken with adventurous publishers. There were many, most of which failed rather quickly because of the high production costs relative to the subscription incomes. A few are still with us. *Scientific American*, in various incarnations, concentrated on practical matters and brief squibs of information and curiosities. The magazine

---

[2]There are several of the original 19th century physics journals that, having passed through a number of transformations, are still published, *e.g.* the Italian journal *Nuovo Cimento*, founded and originally edited by C. Matteucci and F. Piria, and the German *Annalen der Physik*, founded in 1790 and most famously edited by J.C. Poggendorff and P. Drude, covered the physical and mathematical sciences (including chemistry).

[3]To give you an idea of who these editors were, and why the reputations of the journals grew rapidly, it helps to look at a very partial list of names: Volterra, Mittag-Leffler, Liouville, Rayleigh, J.J. Thomson, Tyndall, Drude, Wien, and Planck. All had made – or were at the time of their editorship making – outstanding scientific contributions while also guiding by their selections the direction of entire areas of research. A unique view into this last feature is provided by recent studies of the Planck-Wien editorial correspondence for the *Annalen der Physik* (Hoffman 2005; Pyenson 2005).

we now see is much different from that of the 19th century, which more closely resembled *Popular Mechanics* or *Popular Science* (the *American Scientific Monthly* was more serious). The only one to survive the shakeout of the end of the 19th century to remain in publication, is *Nature*, founded by N. Lockyer in the 1860s with the express intention of presenting novel results, as brief communications, along with news and commentary (Meadows 2008). Again, it was the editors who made the decision on what to publish. The American Association for the Advancement of Science (AAAS, the national counterpart to the British Association) began publishing *Science*, the American survivor of this shake-out period, but it was a membership journal so not subject to the same fortunes as the other commercial ventures.

Another feature of the end of the 19th century was the rise of professional organizations, for instance the American Physical Society (APS), the American Astronomical Society (AAS), and the American Mathematical Society, whose explicit aim was to organize and represent a group of newly minted professional scientists. These were not merely groups of amateurs or enthusiasts, they were, in a real sense, the pioneers of the profession. Recall that before the middle of the century even the professionally employed "scientist" did not exist, much less "physicist".[4] Physics departments, indeed departments and faculties of science, were not founded until the end of the century (although observatories and laboratories did exist separately, as did museums and collections in natural history and botanical gardens that served medical faculties) and the research and educational activity of academic scientists was organized around single professors and institutes. In such an environment, which persisted in Europe far longer than in North America, the various national scientific associations such as the British Association for the Advancement of Science, and the AAAS, mirrored this university-based hierarchy, being organized in sections. Thus, the reports of such meetings contained the physical, biological, and mathematical sciences but did not appear as separate fascicles reporting the workings of specific sections. The societies were usually founded, like the Royal Society, along with dedicated journals (for instance, the *Physical Review*, *cf.* Hartman (1994) for the history of the APS journals).

2.3 The vintage astronomical journals

In astronomy, the observatory publications were internally reviewed before material was published, and these appeared in series exchanged among the observatories internationally. The *Monthly Notices of the Royal Astronomical Society (MNRAS)*, similar to the Royal Society *Proceedings*, presented the communications from the meetings but extended these to submitted articles when submitted through a member (who essentially vouched for its contents). The publication, *The Observatory*, was separate, an originally less technical, non-society publication that nonetheless reported on the proceedings of RAS meetings and often included transcripts of the lectures and debates. Perhaps the most famous

---

[4]It is enough to note that the reigning classic English language monograph on mechanics in the second half of the 19th century was entitled *Treatise on Natural Philosophy* by Thomson (Kelvin) and Tait and retained that title to the end of that century. The Cambridge Philosophical Society centered on mathematical physics and mathematics, yet did not explicitly identify either in its name or journal title.

example of which is the exchange between Chandrasekhar and Eddington on relativistic degenerate gases, see Wali (1991) and the original report, available on the ADS. Another example was F.W. Argelander's *Astronomiche Nachrichten* (hereafter AN), which, while publishing observational particulars (for instance, the announcement of Galle's discovery observations of Neptune), also included theoretical material and short commentaries, but avoided news items or meeting notices.

The landscape changed again with the founding of the *Astrophysical Journal* by G.E. Hale in 1895 and the *Astronomical Journal* by B. Gould and B. Pierce a few years earlier. The *AJ* was an American answer to *AN*, and less so to *MNRAS*, being a venue mainly for observational results, especially in astrometry and celestial mechanics, and reflecting the intended audience. It was important that neither of these new journals required communication by a member of responsible editor, both established an open submission policy, which required some form of refereeing in the absence of "pre-endorsement" . It was paralleled by a journal with a familiar name, but *very* different character, *Astronomy and Astrophysics*, a short-lived journal that explicitly included the new discipline of *astrophysics* in the title as a polemical provocation. It was Hale's intention to promote this new science (see Wright 1966; the details are found in the centennial volume of the AAS and also in the online history at `http://www.aas.org/had/aashistory/`). The journal was not explicitly linked to the society, in part because of the conflict over the naming of the new professional organization – the exclusion of the word "astrophysics" from the organizational designation, see Berendzen (1974) for a particularly detailed study. The observatories continued their own series for a very long time even with these alternate venues since they did not have the limitations of profit and space.[5]

In founding the *ApJ*, Hale made two interesting decisions. One was to assemble an international advisory board, a very striking attempt to legitimize the journal from the start, including European scientists. The second was to emphasize a technique, spectroscopy, that he felt was the defining distinction between the more traditional areas of astronomy and the possible new science, the physical content of the study of celestial objects. A few spectroscopic societies existed or were sections of existing societies, such as the Italian Society of Spectroscopists, but the *ApJ* was an independent body. Since at the time Hale was at the University of Chicago, itself recently founded, he successfully negotiated the founding with the university press and the ApJ became Hale's personal outlet and the university's property. It was not long afterward, however, that Hale moved to California to realize the construction of Mount Wilson Observatory (later, of course, Palomar) but the journal production remained at Chicago with two offices. The editors were Hale at Mount Wilson and Frost, who succeeded him as director of Yerkes. This highlights the unusual nature of the *ApJ*. The editors were at observatories and, as such, controlled their staff's output but also made it simple for papers to be submitted by walking down the hall and handing it to the principal. Refereeing was introduced but not always used, observatory staff had a priority in the publication, accounting, for the predominance of American authors. Multiple publications were tolerated, some of the papers appeared in abbreviated

---

[5]The *ApJ* and *AJ* were founded *before* the community joined in the society-forming mania that dominated the last third of the 19th century, when professional societies were formed to promote the sciences and open the membership to those outside the ranks of the academies.

form in, for instance, the proceedings of the National Academy of Science (prominent examples being the publications by Hubble, van Maanen, Shapley, Hale – who was also one of the NAS founders, surprise!), but these were rapid communications (limited to only a few pages). There were a few conference proceedings, but rather few (the Michelson-Morley experiment conference was one of the few of this sort). Translations of foreign papers were accepted.[6] Laboratory studies were given a pride of place, and series were encouraged (both for the science but also because it was a steady supply of papers on related subjects, some of which were later collected into stand-alone volumes, the most famous instances are Chandrasekhar's monographs[7]). This editorial structure underwent a radical change in the early 1950s. Hale was gone, Frost was retired, W. Adams was at the end of his career, O. Struve had completed his term (also then at Yerkes), and the journal's financial status was perilous (It is the case even though you as authors do not think much of this, that the survival of a journal – even the *arxiv* – requires support). This brought some of the most dramatic changes in the journal structure and, by extension, established the modern form of journals in our field. S. Chandrasekhar was persuaded to take over the sole editorship when W.W. Morgan resigned and the journal was transferred to the AAS – under the encouraging presidency of M. Schwarzschild – which retained copyright, being held for the authors. To ease the financial situation, nominal page charges were introduced (this at a time when the *Physical Review* was paying referees) paid by the authors or by their home institutions as a supplement to the subscriptions. The AAS also provided funding to maintain the editorial offices, and some support also came from the Chicago press.

My reason for this long background discussion is to explain how the modern journals came to be structured as they are. Claude Bertout provides the details for *Astronomy and Astrophysics*, whose history is even more complex, but it is at this stage that you will see what an editor actually does. Chandra was motivated, as Hale had been, by a particular intellectual program, to increase the presence of theory in the journal. As such, by his own output he encouraged a flow of new papers in radiative transfer, hydrodynamics, numerical modeling, stellar dynamics and galactic structure, interstellar medium, cosmology, and new areas of astrophysics such as radio astronomy, while maintaining the well-established areas of solar and stellar physics, laboratory studies, and fundamental properties such as photometry. Wali (1991) reports that he personally refereed about 1/10 of all submissions and had his own papers reviewed in the observatory. *Letters to the editor* were so designated because these were reviewed personally by Chandra without generally being refereed, and comments (short communications) were also included. By the mid-1960s the *Supplements* were added to provide a separate outlet for longer compilations or semi-monographic contributions that were soon to largely replace the

---

[6]This extended a practice from the 19th century of reprint translations, *e.g.* Taylor's *Scientific Memoirs*, and the practice often followed of multiple publications in different languages, *e.g.* Zeeman, Fermi, and others, extending through the 1920s.

[7]An entire issue of the *ApJ* contains what eventually became Chandra's book on stellar dynamics, with a few reprinted papers with, *e.g.* Von Neumann, added. This is an unusual example but it did happen. In some journals now, for instance *ApSS* and *Nuclear Physics*, the volumes are conference proceedings and that also appear as separate books.

observatory publications (but these too had evolved into journals, *e.g. Bulletin of the Astronomical Institutes of the Netherlands (BAN), Bulletin of the Astronomical Institutes of Czechoslovakia (BAC), Astrofisca Norvegica*, and *Acta Astronomica*. Only a few, *Astronomie et Astrophysique), Zeitschrift für Astrophys.*, and still *AJ* followed the same path as the *ApJ*, often with the same centralized editorial structure and refereeing methods. A few new, specialized journals were founded in the 1960s, especially *Solar Physics* and *Icarus*, whose copyright was (and still is) held by commercial houses. As far as I know, among the journals started in this period, only *Astrophysics and Space Science* (the particular project of Z. Kopal, founded as a deliberate competitor to A&A) sought to retain the broad-scope approach of the *ApJ* but with a far looser refereeing structure. The loose board of advisors was replaced by the AAS with an elected publications board that, for the first time, included professional librarians along with working scientists who could advise on bibliographic and archival matters. This last step was vital, for example, to evaluations of new technologies. The board, for the AAS and for *A&A*, operated (and operates) autonomously but doesn't enter into direct editorial matters. These are left to the appointed scientific editors with the result that the functions essentially duplicate the structure envisioned in the 18th century by the *Academié Royale*.[8]

With the American post-WW II technological domination of astrophysics, especially the large observatories (not only Palomar but also Kitt Peak, CTIO, NRAO, and the space program) and large computing capabilities, *ApJ* became the central focus of communication in the field by the end of the 1960s. On retiring at the beginning of the 1970s, Chandra was succeeded by Helmut Abt[9], who took several editorial steps that again changed the face of the journal. The *Letters* was separated from the main journal, explicitly providing more rapid communication in an abbreviated format and under the separate editorship of A. Dalgarno at Harvard. The supplements changed from irregular to regular publication, still under the main journal editor, and the format changed to increase the size of the page and the number of pages that could be published, and gradually increased the frequency of publication (which had been steadily evolving in the preceding years). The refereeing base was extended, including younger scientists, and the journal experimented with alternate forms of publication (videos, included as VHS cassettes, CDs, and ultimately online publication). By the time Abt ended his editorship, in 1998, online publication was the rule, electronic submission had been standardized along with a database for editorial control, and – and this is perhaps the most important innovation by the publication board, a group of scientific editors had been appointed.

---

[8]The International Astronomical Union (IAU) has played an important role in the standardization of the literature since its founding in the early 20th century. While it publishes no journals itself, its presence has been felt throughout the community through its roles in nomenclature and international debate. For instance, specific classification systems, photometric and spectroscopic standards, even the definition of planets – perhaps the most controversial public decision ever taken by the IAU – are reflected throughout the literature.

[9]One of Abt's major contributions, before joining the journal, was to produce – on his own initiative – the first analytic (keyword) index for the *ApJ*. The journal principal copyeditor at the time, Jeanne Hopkins, also compiled the *Glossary of Astronomy and Astrophysics* based on her notes from discussions with Chandra, who wrote the forward.

3 The scientific editorial process now

This now brings us to the present and the point at which I can begin to explain the work of an editor for *A&A*. The history of the journal and its organization and governance are covered by Bertout so I will leave those aside. I only note that a main difference with society publications is the self-propagating nature of the A&A publications board, whose members are selected by the individual national communities and not by election from a membership, a very significant difference. In general, the scientific editors are chosen by invitation or application from the community. For the most prominent journals in our field, those with the longest histories and visibilities, the choice is by an autonomous representative body.[10]

The scientific editors (hereafter SEs, also called "associate editors") are distinguished by their research specialties and assigned papers by the editor in chief (for the main journal and the Letters) in those – and related – areas for oversight during the review process. The number of editors varies enormously among the astronomical journals (*e.g.* there are 14 scientific editors for MN, 18 for ApJ, and 10 for A&A) and even more among those in the physical sciences.[11] When a paper is registered, and the necessary administrative details are attended to, the paper is sent to one of the SEs as a new submission. The paper has already passed a first gate, it is been deemed sane, that is "not crank". You would surely be amused to see some of these but they are not a major concern. The obvious bizarre papers are returned to the authors as inappropriate for the journal with a civilized cover letter (not that this excludes the possibility of threats and appeals by the offended authors). But even a paper that appears to be of marginal relevance to the scope of the journal are now sent to the SEs for comment before a decision is made on whether to send it out for review.

3.1 The editor

The first SE review is a step that needs some explanation. It is not necessarily a judgment by the SE of the correctness of the result, although that comes into play occasionally. Rather, in a world filled with journals that may be more specialized (prestige aside, that is not the point of this decision), papers may be returned because they are better suited to some other venue. This can reflect the evolution of the field – there are specialized journals in almost any area, from general relativity to celestial mechanics to radiative

---

[10]Some journals are, however, almost "personal projects" of the editors – who are often the founders – but in our field there are few of these. There are also some journals founded by the initiative of commercial publishers who decide themselves who to invite. In some cases the editors are nearly full time, salaried employees. In other cases there may be a unique editor.

[11]It is important to note that the astrophysical journals are not subdivided into parts such as, *e.g. Physica*, *Physical Review*, or *Journal of Geophysical Research*. Instead, although there are more specialized journals that address readership in subfields, such as *Journal of Quantitative Spectroscopy and Radiative Transfer*, *Icarus*, *Solar Physics*, and *Celestial Mechanics and Dynamical Astronomy*, the principal journals have chosen to present a more unified perspective and accept papers in a very broad swathe of the research subdisciplines. In this, we are unusual. I know only one field, *Journal of Fluid Mechanics*, that maintains the same very wide scope by deliberate design, similar to Hale's, by the founding editor, G. Batchelor. Batchelor also left a remarkable legacy, his editorial reminscences (Batchelor 1981).

transfer – and given the competition for journal space it is better to send back a paper that will find an audience elsewhere if it passes muster, than overly limit the access of the journal to more relevant papers. More on this point later. There is another, even more important consideration: every paper that goes out for review takes time, freely given as a service to the community, from a working scientist and this is a precious resource that should not, and cannot, be abused. There may seem to be too many astronomers in the world, when bureaucrats tally up their budgets, but the truth is that there are not that many and there are a number of journals, and journal editors, vying for their attention. Every paper under review for the journals in our field is read by the editor and by the referee, often several times along with the responses, and this requires a huge dedication of time and attention by all concerned. An author may think it is something just dashed off, if you have never written a review, but a single paper may require days, or weeks, of time by those who will never receive more than an acknowledgment (at most) or remain anonymous to the community, known only to the editors (at least). That the reviews are anonymous, except when explicitly waived (see below), may seem like a way of hiding behind a comfortable mask while delivering a killing stroke to a paper but the editor is there to insure that this does not happen and this is the greatest benefit to the community of the SE structure.

3.2   The referee

The referee has, then, several responsibilities. One is to the community. As a privileged reader of a submission, both as evaluator and as a fellow scientist, the referee sees the paper – unless it is been placed on the *arxiv* at or even before submission – before anyone else except the editors and is thus party to information not otherwise available. This can put a reviewer in a delicate position of conflict of interest, if the research is too close to her/his own work, but it is almost always (and those few exceptions will be discussed further in a moment) signaled at the time of the request. The process begins with the editor sending a copy of the abstract to the prospective reviewer, and if properly written, it will contain more than enough information to permit a potential reader to decide whether to accept the task. But this is also privileged information. We know well that a colleague, or competitor, can extract a wealth of information about a result from just a few key words. An abstract that is properly written for later introduction of the paper to the readership will also convey enough information to an unethical reader to jump the process and try to publish first. One of the most impressive characteristics of the astronomical community that this rarely happens. It is more likely that someone will see a talk at a meeting.[12]

How the review process starts and is monitored and mediated illustrates the centrality of the editorial board in the process and highlights the importance of *scientific* (as opposed to administrative) editorial control of the refereeing process, The editor in chief receives

---

[12] An anecdote is in order here to emphasize the difference between how referees *do not* behave. Recently, "preliminary" data presented in a slide at a meeting has been photographed by a listener using a cellphone and then sent directly to competing groups, with a paper being submitted before the end of the meeting discussing data that has yet to be submitted. One of the editorial roles is also to have a sense of when this is happening, but more on that later.

the paper for consideration when all bureaucratic requirements have been satisfied and, on reading the paper, selects the appropriate SE for overseeing the evaluation. Some journals have a schedule for distributing the papers to the editors, for others it is a continuous process. The arrival rate for papers is certainly *not* a random variable, since there are strong correlations with the outside world of deadlines (for, *e.g.* observing time proposals, grant applications, fellowships, theses, vacation periods) and these affect all journals. There is also a strong "Monday Morning effect", when papers that have arrived after close of business on Friday may have to wait for assignment depending on how the editorial staff is scheduled. Again, this differs between journals but also affects later contacts with possible reviewers. So assuming all goes smoothly, a paper may arrive in the mailbox of an SE up to a few days after the authors have been told that the process has been started.

The criteria for selecting a referee are flexible, and this is where the process is most vulnerable – at least in principle – to the whims of an editor. The wrong choice of a reviewer colors the whole process, but it is the role of the editor to have sufficient background in the field to decide who can both evaluate a paper on its scientific merits and provide an external reading. And for this reason the editor is always visible in the process, the one who ultimately takes responsibility. The choice is made on several different levels. Along with the paper, the authors can state a preference for which SE they think would be the appropriate choice for overseeing the evaluation. They can also suggest potential referees, or signal those who they prefer to see excluded.

This information is considered at all stages but there should be explicit reasons for avoiding an individual or group and/or additional information for those in the preferred category.[13] Close colleagues are excluded, or at least close collaborators, usually based on publication history. Competitors are also excluded, although more on this point in a moment. The narrowness of some subfields can make this step particularly difficult, especially for large collaborative papers that can virtually preempt a whole field. But at this point, the first editorial reading of the contents of the paper is important because there may be some outside the precise area of study who are perfectly capable of evaluating the paper without being parties to the various controversies. The bibliography is important in this regard, it may indicate some of the connections that the editor does not know, or indicate who might be outside the group but not compromised by rivalry. The ADS is another important source. Tracing back a citation history for a possible referee can reveal connections, possible biases, or other problems. It can also provide useful additional context to the editor.[14] Depending on the number of papers being dealt with, an individual editor may take some days to begin the contacts (an invisible step, the authors see only the date of registration of the paper).

There are also more than one editor for the journal and even SEs can collide, when the ideal reviewer for one paper may already have recently accepted the request of another

---

[13] It has happened that authors have suggested collaborators as reviewers, for instance. This is *not* a very good idea: not only does it look ridiculous, it also indicates an irresponsible approach to the publication process. In other words, be thoughtful when making any reviewer suggestions.

[14] This is neither an instantaneous process, nor can it be automated (well, it can be but I am reminded of the play on words of a sign in a laundry "we do not tear your clothes by automated machinery, we do it carefully by hand").

editor to take a paper, or may have been asked recently enough that a new request would be an abuse. Some colleagues, dissatisfied with some previous experience with an editor or a journal may systematically remove themselves – implicitly or explicitly – from the process, further reducing the available pool (although this does not happen often it has and does occur).[15] At times, even if unable to take a paper, a contacted reader will suggest an alternate referee. Like the author's suggestions, this is taken under advisement but not always followed. It is, however, an excellent way to get younger scientists into the refereeing pool.

An important distinction in both the selection and review process between the Letters and the main journal is the criterion for acceptance. This is similar for most of the astrophysical journals. A Letter must be both timely (read "urgent") and particularly interesting. The latter is, of course, a value judgment but it can be substantiated and, indeed, must be in any evaluation. For some journals, the paper may be more speculative, perhaps even on the borderline of being "far out", but it must not contain any obvious errors. The timescale for evaluation, usually about half or less than the nominal time allowed for a main journal submission, means that the process is somewhat looser but no less serious. A Research Note requires less contextualizing than a regular paper but does not have the urgency of a Letter. Actually, in this age of instantaneous communication of preprints, the Letters have an odor of anachronism. They continue to be regarded as, somehow, more prestigious than the main journal, a hang-over from earlier times.[16]

When the person has been selected, the contact process begins. This can be long and, from the outside, may seem very inefficient. To an author consulting the submission system (in our case MMS), to see a list of contacts of contacts (by number only, *i.e.* referee 1, referee 2, etc.) doesn't necessarily inspire confidence. It may appear that the editor is "scraping the bottom" to find a reviewer. But we are all busy and to review a paper takes time. There are always those who do not respond the first time, or the second (and editors reserve a Dantean ending for them), but other factors can enter. Spam filters can block messages without sending notification to the journal.[17] Assuming the

---

[15]Also, keep in mind the number of submissions per year to the principal journals, including those in other areas of physics, informatics, and even engineering, for which any scientist may be asked to review. There are literally tens of thousands of such requests in circulation each year, an almost terrifying volume of papers! It should, therefore, not be a surprise that at times a paper can take a number of requests, to very qualified scientists, to finally end up with an appropriate reader who *is* available.

[16]This is particularly true for the physics community that regards the *Physical Review Letters* as the place of choice for "important" results even though any paper submitted there can also be placed online and the statistics of acceptance are quite skewed by the evaluation process. Let me explain. A paper can easily be rejected after a single pass if there are two (or more) reviewers, as happens with some journals. The revised paper, resubmitted, may make it through easily yet the statistics for acceptance – the impact factor – will be distorted by the process. Main journal submissions usually go through at least one, sometimes two rounds. An unscrupulous journal could easily increase its statistical standing by changing this procedure. Fortunately, most journals are not yet succumbing to this temptation.

[17]This has been a particular nuisance, or worse, in the past few years. Some addresses have been flagged without notification. For example, mindlessly applying the rule that uppercase letters with numbers, or words such as "urgent", may indicate spam or virus-laden mail, some sites have blocked the journal addresses (*e.g.* AA), necessitating workarounds. The database a journal maintains of email addresses is only as up to date as the last contact and even those entries may contain errors, from changes of domains or changes of names. These all slow the process.

potential reviewer does reply, the response may be that the request comes at a bad time, or that the paper is too far from her/his current interests, or even that there is a possible conflict of interest in the review. The she/he may suggest someone else as a better suited for the paper. Remember, there are a number of journals and as a round number we have about 10^4 potential reviewers, around 3000 papers per journal per year, and about a dozen journals just in astronomy (not to mention broader and more specialized physics, numerical methods, and engineering journals) so the number of requests is potentially large for the more visible scientists and the probability high that more than one request will arrive at any time (an interesting exercise in Poisson statistics). The review is *not* a random process but there *is* a significant stochastic element at the beginning.

There is also the occasional author – fortunately a very small minority – who simply refuses to participate in the process as a referee. This is not only unfair (well, perhaps not considering what might result in the report) but also irresponsible. Authors takes a referee's time, whether they want to be peer-reviewed or not. By the submission they explicitly accept the process. It is a singularly uncollegial act to deny the same journal some effort to provide advice in reciprocity.[18] Some journals explicitly recognize referees. For instance, the AGU and APS have special awards for reviewers who, in the view of the editors of the respective journals, have been particularly noteworthy and conscientious, *Icarus* publishes a list of the referees at the end of the year, and it is not infrequent to see in a *curriculum vitae* a list of journals for which a scientist has been asked to serve as referee. It is a special pleasure for an editor when a reviewer not only enthusiastically accepts the task but, after a careful report, is happy for having been asked. It *is* an honor to be asked to serve as a referee: those who receive a request for a report are asked based on their expertise and publications, not simply because they happen to be on a list or in the bibliography, that's only the start to jog an editor's mind and to keep the pool of reviewers constantly renewing and open.

A potential reviewer should immediately indicate possible conflicts of interest, for example competing research unknown to the editors, that might compromise the process. It also happens that, in refusing a request, a contact may explain biasing predispositions regarding related work by the authors. These notifications are taken quite seriously and are basic examples of ethical behavior.

Assuming all goes well, a paper is placed at the first request, but there may still be a delay waiting for the paper to be retrieved. This is flagged, after some preset waiting period, by the database itself based on experience of the editorial office. We allow 10 days before a flag is raised. A few reminders at this stage may show that the review cannot continue and a new round of refereeing requests begins. Otherwise, when the flag drops and the review is assumed to have started, there is a wait of between 3 to 4 weeks for the first report, for the Letters it is one to three. That is, again, if all goes well. The statistics show that the average reviewer will need one reminder, send before the journal's internal

---

[18]The referee's history is also indicated in some journal databases, *e.g.* how many times he or she has been contacted during some calendar period and whether the reviewer accepted the request or not. This is also an important reason why it is a requirement that the submitting author signal whether the paper had been previously submitted elsewhere. This helps avoid multiple reviews for separate journals by the same reader, something that really happens and can lead to interesting results.

deadline but after the period usually anticipated for the review. Some referees have been known to disappear at this stage, a serious, unexpected problem.

3.3   The referee report

The best referee report is one that serves both the authors and the editors. For the authors, it provides an independent, critical reading of the paper, suggestions for improvements or expansions, and flags weak assumptions or sloppy or nebulous presentations. For the journal, it signals where there may be serious problems (for instance, plagiarism, see below), improper citation of – or misrepresentation of – the existing literature (and this abuse goes far beyond issues of credit for discovery, etc.), redundancies (including incremental extension of previously published results of insufficient significance to warrant separate publication), or basic errors. The report should be detailed, grave problems must be flagged and extensively discussed. The report is read by the editor before being sent to the authors, and this includes possible correspondence with the reviewer to clarify points or to request extensions. It is possible that the editor her/himself will add comments that must be addressed in the revisions and responses. If the review is uncivil, or too strong, or the language is poor or ambiguous, it may be edited before being sent out. If it is ambiguous, this is always cleared up before it is forwarded to the authors.[19]

The report itself is structured, at least in part, by a set of standard questions to be addressed by the reviewer.[20] These are similar across the spectrum of journals, following the historical lines I outlined earlier, and are to be explicated, in detail, in the review. The best reviews can extend for many pages, not that length alone is a criterion. If the depth of reading appears superficial, it may be sent out but with a comment by the editor; in the case of a negative review this may lead to an immediate offer of a second referee (if the correspondence with the referee has not produced a revised version). Since authors may become referees, it is appropriate to insert a few additional comments.

A central issue, "why should this paper be published", is stated rather baldly but addresses a key issue. If the material is merely incremental – what has been sarcastically called a "minimum publishable unit" – or contains largely uninterpreted data (the sort of paper that, in years past, went to the *A&A Supplements*, now discontinued), it may be deemed inappropriate for publication. A separate question is whether papers in a series can be combined or if the series is worth continuing. For this reason, it may be appropriate to have a single reviewer take more than one such paper (if they arrive in tandem) or

---

[19]The report may also, depending on the journal and editor, be check for possible sexist – or "gendered" – language, in part because the report could be unintentionally offensive (this is one of the cultural shifts from 20 years ago and while it can lead to the occasional awkward construction it has great advantages in preventing misunderstandings). This does not apply to the confidential comments that may precede the report, but even there it may be indicative of problems that need to be discussed before a report is sent.

[20]We have an unusual one in astronomy, whether the objects are denoted according to IAU nomenclature. This is frequently confusing, perhaps even amusing, to those outside the field but the importance of SIMBAD and the enormous importance of the CDS, as Laurent Cambresy discusses in his chapter, makes this an essential check of the manuscript. Think of what chaos can be created by changing the name of an object or using a nonstandard format for a catalog. It may seem an annoyance but without this standardization it may be impossible for work to be checked or may even lead to incorrect search results or citation problems. We ae hardly the only field to have this problem, think of chemistry, geology, and biology.

to line up the same reviewer for future submissions, should the report show that this will be useful. It is always a potential problem when different editors handle papers in the same series without coordinating the reviews. There are various levels of revision. For cases of truly minor changes, mainly presentation (*i.e.* tables or figures that require restructuring), the referee exits the process. If, instead, the changes and suggestions are basic to the science but not especially critical, the paper will still go back to the referee in revision but it is virtually at the end of the process and such second reviews are often a formality. At times, however, even these may take extra time if the authors have not properly addressed the criticisms. This does not mean they have not changed the paper, but ignoring a comment is more than bad form, it is incorrect. The comments are included for a reason, and the reviewer and editor deserve explanations if a critique is deemed irrelevant in the authors' judgment. If the criticisms are particularly serious, perhaps fundamental, then the authors may receive the suggestion that the paper be withdrawn and reworked – not merely rewritten – and resubmitted as a new paper. In that case, the paper will enter the queue at the top and may go to a different scientific editor. If the authors choose instead to continue the review, then a very substantial effort is likely to be required and the review may take as long for the revised version as for the original.

3.4   The review process as a queue

The load of an SE varies not only with the day of the week and the season, but also with the hardest of all variables to predict, the return time for a revised manuscript. This is the least controllable part of the refereeing process and, as Bertout's (A&A) and Abt's (ApJ) studies have shown, accounts for the bulk of the time required before a final decision is made on a paper. The median time is relatively short (see Bertout's discussion) but the process is far from Poisson and there is a "fat tail" distribution in the waiting time. The refereeing process is actually a queue, for A&A it is a multiple server, Markovian process.[21] That means there are several servers (a word that carries various connotations, in this case it is the editor) with the papers being arriving units with some temporal distribution on first entry. The service time, the time required to exit the queue, is much more complicated than the usual theoretical problem. There are several components to the waiting time:

1. Arrival rate: papers enter the queue when received at the editorial office through electronic submission, regardless of the journal. The rate is, however, not constant. There are trends through a year (for A&A in the period 2006–2008 – this is discussed in more depth in Bertout's chapter – this varies by up to 50%, with minima

---

[21]I am introducing this way of viewing the process for two reasons. Astronomers know little about queues beyond banks, supermarkets, and traffic jams and it is a very lovely research area that could be exploited in astrophysics. The other reason, more germane, is that it provides a synthetic tool for understanding any process that depends on flows within a service structure.

generally in Jan/Feb and Aug/Sep).[22] This has a ripple effect through the review process since it determines the load of an SE.

2. Assignment to an SE: almost a deterministic process but may depend on the current load of the appropriate editors;

3. The initial search for a referee: normally of order $n_c \Delta t_c$, where Δt_c is a mean waiting time for a response, allowing for a second contact if appropriate, hence n_c, the number of contacts, includes multiple requests to single reviewers;

4. retrieval time: this is usually short, but can be comparable to the initial search, $1 \leq \Delta t_r \leq 15$ days. This can, and likely will, be, be reduced and depends on the part of the journal (*i.e. Letters* or *Main Journal*). It is a carefully checked step and reminder letters and other contacts are used to prod the start;

5. The review time for first pass: Here a few $\leq \Delta t_0 \leq \Delta_{max}$ where the upper limit is flexible but generally 4 weeks. This depends on criteria set by the individual journals and sections, from a few weeks to a few months (not in astrophysical journals but common in mathematics and history/philosophy of science, and even informatics);

6. The processing time for the first report: this is complex, it may include correspondence between the editor and referee on specific questions but in general it is also short, usually only a few days);

7. The waiting time for first revision: $\Delta t_1 \geq 1$ day: There is an effective upper bound that depends on the journal, when a revision has been awaited for so long that it is effectively a new review because the referee (and/or the field) have changed. The paper may be so radically altered that it's a virtually new submission if the return time is too long. These files are closed by the editor, who periodically culls the papers in this stage depending on date of last action but it can be reopened. This is not a rejection *per se* but amounts to one *de facto*. The distribution of Δt_1 is dominated by correlations and fat tail distributions being closer to a power law than a Gaussian;

8. The return time to the referee: This depends strongly on the time of year and may be longer than the first contact simply because there is no "right of refusal" at this stage, if the reviewer is not available the manuscript sits in a holding pattern);

---

[22] There are many good sociological reasons for this variation. For instance, observatories have fixed dates for proposal submission, as do many granting agencies. These often cause authors to push the submission dates for papers to demonstrate their productivity. Periodic meetings – the IAU General Assembly, or the AAS winter and summer meetings, or the AGU winter meeting – for instance, can have the opposite effect, temporarily removing both authors and referees from the pool due to the main meeting and attendant symposia. The start of the academic year, and its end, are also perturbations for both parties. There are, of course, the particular months so well known in Europe of July and August when the world beats a path to vacation, even astronomers, and there is always the start of the new year.

9. Report on the revised version, Δt_2, which generally has a similar form but truncated limits relative to Δt_0. One note: a revised paper may require as much, perhaps even more, time to referee than a first submission, the authors can help significantly by carefully detailing their revisions in the manuscript and thoughtfully and extensively responding to the report (we request keying to the relevant items) in their cover letter (which is forwarded to the referee along with the manuscript). One reason, alas, for a delay is that the responses are not always in accord with the text of the changes, or the reasons included in the cover letter are so important that they should be included in the paper;

10. This process is iterated, usually at least once but perhaps more times, see below for a discussion of how impasses are resolved.

There is a parallel queue for *Letters*. Most journals, if they have such a section, create an independent editorial structure to oversee the process, including SEs and a separate referee pool. This review procedure can actually set the journal in competition with itself if the possibility exists to transfer a paper, at the level of the editors-in-chief, from one part to another. At A&A, a radically different structure has evolved with the SEs overseeing both parts, thus allowing seamless transfer between the two. Such decisions are made according to well-defined, although flexible, criteria.

The most difficult step is the wait for the return of the first revision. The timing is important and uncontrollable. The load of an editor can vary but for A&A it consists of some number of manuscripts to be placed, some number awaiting a referee, and those in active review, and can be between 100 and 200 at any time. Some fraction will be in first review, some in second, some in third, some in second refereeing, so you get the picture. Some further number will have been recommended for acceptance and are on hold, some will be with the language editors, others awaiting publication. That the process takes a long time can, I hope, now be easily understood.

3.5 Second referee

This is left open-ended because there is a possible branching factor. Let me explain. Some journals use two or even three reviewers and, based on a weighting scheme, determine if the paper is to be refused at first review. If a revision is allowed, it may still lead to a rejection, or it may be rejected up front at the revision stage by the editor. In single-referee review processes, should the editor determine that an impasse has occurred, or if the referee signals an unwillingness to continue, the authors may be permitted to request a second reviewer – with justification and appropriate revision of the paper before submission. It can also happen that there is clearly no point because the referee has simply refused to reconsider the manuscript after the first review. The second reviewer can be independent or "informed", the first being a completely new review of an obviously revised paper (at least with knowledge that the paper was already through one review cycle) while the second means the responses – the history of the review cycle – has been supplied. There's a sort of third option, sometimes requested by the second referee *or* offered by the editor: at the end of the review of the revised paper, once the report has been received at the journal, the referee may get to see the original comments as a calibration of her/his report. This is, of course, one of the reasons the editor's oversight is so important – the editor sees all correspondence and can judge whether the report of the first reviewer is really to be ignored, or that of the second, or whether there are common points that should be signaled in the cover letter to the authors.

4   Particular issues

4.1   *The availability of information: from* Jahrsberichte *to the* Astronomical Data System *(ADS)*

The world is awash in scientific publications. There are hundreds of conferences, each of which produces proceedings, and hundreds of thousands of pages published each year. To say it is folly to follow the literature at such a detailed level is an understatement (more on this when we discuss citations). But since the days of the *Phil Trans.*, there have been aids for the bewildered and overwhelmed. These are the abstracts published by dedicated reviewers of the journals since the 18th century. A particularly noteworthy example is the review journal *Astronomischer Jahrsberichte*, which published reviews of the astronomical literature for many decades, which was a review abstracting service, still extant in the mathematics community as *Mathematical Reviews* assigned papers for review to a large pool of "post publication referees". These were/are more than abstracts; they are usually brief (or not so brief) critiques of the paper. The *Astronomy and Astrophysics Abstracts* was different. As the forerunner of the Astronomical Data System, it was a compilation of the abstracts of papers. Its principal advantage over merely browsing the shelves was an extensive taxonomic system that placed cross-referenced keyword codes alongside the abstract to render the paper more accessible (as Abt did for the *ApJ*). This was the underlying informatics architecture adopted by the ADS, that permits searches by keyword – not merely title and author – but with the addition possible only with the development of sophisticated database manipulation, to link papers at many search levels, also linking the reference lists (as done also by *Science Citation Index* and related databases) but also connecting to the paper itself and to data and other material relevant to the study. This is covered further in Laurent Cambresy's chapter on the workings of the Centre de Données Astronomiques de Strasbourg (CDS) in general and *SIMBAD* in particular, so you are referred to that discussion. The journals still maintain their own categories and keywords, subject to revision over time by editorial review, and the authors can choose the subsections for those journals that explicitly have them (and, collaterally, to state a preference for the choice of scientific editor).

4.2   On anonymity

The referee process appears lopsided. The referee knows the author(s), while the author(s) generally do not know who the referee is. You know from grant and observing time evaluations, and job applications, that this is very discomforting. But it has some value in keeping the process impersonal since the mediation, and moderation of the discourse, is managed by the SE.[23] Some journals use double-blind reviewing, where both sides are anonymous. This is used frequently outside of the sciences so let me explain how it works (since it is possible you have never seen the process in, say, history).

The main difference between the editorial role I have been describing and the editor of, say, a journal in the history of science is to insure complete anonymity. This is why you see such stilted language in some disciplines, the impersonal is used to hide the author's identity and parts of the text may even be "censored" (read "blacked out") to insure that self-citation does not reveal anything (*e.g.* instead of saying "as we showed in Blop *et al.* (2008)" this would read "as Blop *et al.* (2008) have shown", a minor difference in wording but one that signals a very different mindset. When in some areas of research the authors are many and the individual products are relatively few, this works. We, on the other hand, are more accustomed to building a context by citing previous work by ourselves as well as others and part of the value of referencing is how it can – used assiduously – keep a paper short but still inclusive. In grant reviews this is obviously impossible since the track record of an individual or group enters the weighting scheme for estimating the likelihood of future success.[24] It is not, in astrophysics, all that hard to narrow the list of possible authors or possible reviewers given the citations. So in a better world, it might be reasonable to go this way but in practice it is really the editors' role that makes this less necessary.

The opposite choice would be completely open review. This is reasonable and democratic but not without possible abuses. The public posting of reviews of papers, as some have suggested for open access, is based on a market analogy that can seriously damage a paper and/or its authors and might well lead to a mere degeneration toward the mediocre. By this I mean that if both parties are airing their views in public, both parties may be cautious for careerist reasons – their own and/or their students and/or collaborators – since the fear of revenge is a very human feeling and not easily overcome. Moreover, in a litigious society, a feature of almost all western countries, the consequences may be even more serious.[25]

Open reviewing is, therefore, now left as a choice for the referee. Correspondence are requested for the journal files, and the information of the editor in charge of the review, which may be critically important at acceptance. This can be very useful to the referee, since it makes the use of direct discourse far simpler. But it can – fortunately very rarely – lead to "referee abuse", with the authors making the reviewer a *de facto* collaborator without even an acknowledgment (yes, it happens). But again, it is rare and editors often query

---

[23] As a procedural note, the scientific journals have, by and large, adopted the "Nobel Prize rule", releasing their referee reports for scholarly use after 50 years. Some journals make this an explicit agreement, signed by the authors and known to the referees. For others it is implicit.

[24] This is an example of the Matthew Principle, explicated by Robert Merton (1968) and elaborated by generations of sociologists of science but it has its value, along with bibliometric studies. Those who have been successful, or have been in the forefront of a research program, are often in a better position to make future advances or make reliable statements when presenting their results. An illustrative application of this to citations is found in Brown (2004).

[25] To illustrate this point, a simple review of the costs of scientific journals published in *Physics Today* in 1983 led to a more than decade-long international lawsuit by a presumably aggrieved publisher resulting in enormous costs, personal tragedy, and an inconclusive and generally unsatisfactory end.

the reviewer who has selected this option to be certain (the choice of waiving anonymity can actually be an error).[26]

4.3 Referee ethics: the norm and the breech

On accepting the task of reviewing a manuscript, the referee accepts the implicit (sometimes *explicit*) rules governing the review. In the days before electronic preprint servers, the reviewer was in a uniquely privileged position, as the person who would see and critique a work before it was offered to the community at large. Even with the alternative means now available to permit access to the manuscript, the referee is ethically bound to behave with the same discretion and honesty, not revealing the contents of the work to colleagues and considering the manuscript a confidential communication. She or he must not make any use of the material, exactly as in the review of a proposal for observing time or a grant, or in any manner deliberately interfere with the review process. To violate this is more than a serious issue: it is a capital offense against the only thing that keeps the scientific enterprise functioning in the open, it is a breech of trust. The ethic of scientists is to be open, and even if the reviewer is anonymous it is perhaps the most important role of the editor to insure that there is no transgression. If a conflict of interest arises during the review, it is the obligation of the reviewer to immediately notify the journal. If this is clear at the start (sometimes it is not obvious from the abstract and only becomes evident once the actual manuscript is in hand) to not send a notification to the editor is dishonest. For this reason, among others, it is best to at least quickly give a quick pass through the paper to identify possible problems. Most of the time, in fact the overwhelming majority of cases, this is exactly how it happens and the problems (which are few in any year for any editor) can be dealt with quickly. But there are some who, by ignorance or design, seek to delay or even kill a paper for their own benefit. I repeat, this is a rare event, but "rare" is – alas – not "unknown". This does indescribable damage at many levels. It weakens the faith in the efficiency and transparency of the review process. It reinforces the prejudice that the publication of a scientific paper is a sort of struggle against the Dark Side. A few examples are the reminiscences of Parker (1997), the history of Einstein's dealings with *Phys. Rev.* by Kennerfick (2005), and the particularly caustic remarks by Eggen (1993), and also Cole et al. (1981) for the view from sociologists. There are places reserved in *Inferno* for this sort of person.

4.4 Citation: the editorial side of the issue

The footnote was one of the most important inventions in scholarship, but it was preceded by the gloss – the marginal comments inserted by readers of manuscripts in the centuries before printing – that often not only expanded on the text but indicated the sources in a

---

[26]There is one other alternative used by the *AGU* journals. Referees are openly acknowledged, along with the indication of the responsible editor, at the end of the *published* paper. This is by agreement with the reviewers, some of whom nonetheless prefer anonymity. But these papers receive more than one referee and, in the case of *Reviews of Geophysics*, often more than one depending on the interdisciplinarity of the article. The editors are also solicited for nominations for special recognition of notable reviewers by the society at the end of the year, published in *EOS*.

way not done in the text. It was not unusual in the Middle Ages for texts to contain pieces lifted from other works without citation and it is still one of the tasks of scholars to note the sources of the material (which can often lead to new historical insights). This was no mere token of authorship, but was instead a way of contextualizing the material. A citation meant, and means, the full source, with all its individual peculiarities and content, can be traced back without the need for the author(s) to duplicate the information. There is, of course, the other reason, that a living author explicitly acknowledges an intellectual dependence and debt with a citation. It also means that the work has been read and understood (well, at least read). One not-so-humorous definition of a *classic* is a work that is frequently cited but rarely read (see *e.g.* Evans 2008). How many of you, for example, have gone back to the original papers on general relativity, quantum mechanics, stellar structure? Even if these have now passed into history, suffering the fate of "obliteration" – another phenomenon identified by Robert Merton in the 1940s, when a work becomes common intellectual property and the original work no longer needs citation – it is nonetheless assumed that cited it has been read! The same is true for any more modern paper.

Review papers have a special place in this discussion. Its scope is often to assemble and order a vast, dispersed literature. But in the process, a single review paper can effectively "kill" the literature it attempts to synthesize. The most cited papers are, generally, those in *Annual Reviews* and *Reviews of Modern Physics*. Of course, one reason is their singular utility, points of reference from which a study can depart without recreating the whole of the earlier work in a field. But at the same time, the review can be cited (the proverbial "see ··· and references therein") as the unique precedent for any study without taking the time to read either it or any of the literature on which it is based. In an important study, Brown (2004) discusses the diffusion of such citations in the literature based on internet searches.

Bibliographic "purity" is very hard to achieve but it is as important as nomenclature. The error rate in bibliographic databases is improving but it cannot reach perfection so imperative that authors do their utmost to play their part. We cannot rely on the librarians and it just makes their life harder when we producers of the literature are sloppy. A wrong citation isn't just a matter of incorrect formatting. It can create a work that doesn't exist. A phantom paper is a permanent problem. I give a simple example. Last year I was checking the citation of a classic paper, the founding paper on the Kelvin-Helmholtz instability. The text and bibliography referred to it as Helmholtz & Monats. But the actual paper is by Helmholtz alone so I checked the ADS and then started a sort of traceback through the physics journals. The paper is a phantom that's propagated through the literature. It *should* be Helmholtz alone – *Monats* is actually the abbreviation for *Monatsblatt* (Monthly). As a result, there is now a permanent record of a well cited author who was born and died in a single year and produced only one notable paper in his short life. To coin a phase, I will refer to this as the "Lt. Kije" phenomenon.[27]

---

[27]For those who love music, the suite by Prokofiev by this same name will be familiar. It is based on the story by Yuri Tynyanov (born 1894, died 1943) about a mythical Russian army officer created by a company of soldiers whose birth, life, and death are fabricated to fool bureaucrats.

4.5   Series

How do you decide whether a paper should be published singly or as a series? As publication and citation counts become progressively more important for careers, so has the proliferation of series. These may be implicit – very similar, cross-referenced papers published in tandem or successively – or explicit, with division into labeled "parts". This is not new, of course, since whole books have been constructed from some of the historically more famous examples. The Dover reprint volume *Physical Processes in Gaeous Nebulae*, edited by Menzel in the 1960s, is largely composed of the *ApJ* series of the same title that ran to 15 parts from 1937 to 1945 (Aller 1999). Chandrasekhar's series "On the Radiative Equilibrium of a Stellar Atmosphere" had 24 parts in the *ApJ* from 1944 through 1948 and was finally condensed in his monograph *Radiative Transfer* in 1950. The series "Wavelength Dependence of Polarization" appeared in 40 parts in *AJ* from 1960 through 1980, many dealing with single objects, which spawned volumes in the Arizona series of astrophysical review volumes. The alternative would have been to give each a separate title, perhaps less obviously sequentially connected, and use more space in each recontextualizing the work.

A few words of advice. First, the decision should be based on two things: readability and scope of the work. If too many related but disparate topics are presented in a single paper, the reader may well lose the thread. It may be better to split the paper along some natural lines, but this should be done at the start. In such cases, the parts should be "natural" units that really follow each other and build a result. This does not mean, for instance, that multiple objects or case studies necessarily should be – or even deserve to be – separated. On the contrary, the case may be stronger if statistically meaningful samples, or related objects, are presented together.

The decision should be scientific, not stylistic. It can be embarrassing to have a Part 1 published alone, and if dealt with by different editors and/or referees it can happen that one or more of the subsequent papers do not make it through the process or is combined, leading to a peculiar sequencing (*i.e.* when, say, Paper 4 is accepted but not Paper 3). Such things happen. It is, therefore, vitally important that the papers do not "refer forward" – that is, one paper does not leave *essential* details to be explained in a future submission. A review cannot be based on a promise of details of an essential item or draw conclusions based on results that have not yet been either submitted or completed. This is true for *any* paper: material cited as "in preparation" is unacceptable if it is at all important for understanding and judging the results in a manuscript. It may limit the choice of publication venue since some journals are reluctant to continue the numbering of papers in a series for which some have appeared in other journals. It may also lock the series into review by a small pool of referees and will usually be assigned to the same editor (although it happens that sometimes several editors handle the series but may consult among themselves).

Although little studied, Schultz (2010) provides a snapshot from the editorial side (as editor in chief of *Monthly Weather Reviews*, one of the American Meteorological Society journals). If submitted together, for instance, as Parts 1 and 2, his study showed an acceptance rate essentially the same as a single paper. If submitted serially with some time between the arrival of the two parts (*e.g.* if one has already been published), the second has a higher than average acceptance rate. This should not be misunderstood as advice on how to manipulate the review process. Quite the contrary. The higher acceptance rate (about 3/4 compared to 2/3 for the average) is because the multiple parts are linked well and are logically sequential units. When the papers are submitted together, there is also the chance that – sent to a single referee – they will end up combined (although Schultz found, from a ten year sample, that the acceptance rate for the two together was the same as single papers). His final advice is to "write manuscripts that are sensibly [sic] independent of each other, make minimal reference to unsubmitted manuscripts, and have sufficient and substantiated scientific content within each manuscript", all common sense notions but more often ignored than followed.

4.6 Multiple submissions

This can be briefly stated: the submission of a paper to more than one journal is unethical and prohibited. Period. Such a transgression can lead to an author being banned, perhaps permanently.

4.7 Plagiarism: intellectual property and authorship

To explain the problem and disease – not too strong a word – of plagiarism requires a context. The notion of intellectual property is fundamental to the scholarly enterprise and as old as the scientific societies and journals. At its core lies a very human feature of what we do, wanting *our* contributions be recognized by our peers as original, perhaps even valuable efforts in the advancement of science. This is very different from ownership: what we produce, think, and communicate is given openly and freely to the community at large. It hardly even hints at a capitalist idea of property leased or rights sold. The idea of copyright was born in the 18th and 19th century as a way of protecting the expropriation, in part wholesale, of an individual's work (see Rose 1993 for a superb history). Our compensation, if there is any way of explaining it in such terms, is that we have attached to our names (in a singular or collective sense) the air of reliability and novelty, of creativity and knowledge. These are abstractions when so categorized but they are the very personal reason most of us become scientists, to participate as equals in an international community in a historical enterprise. We all know that the literature is ultimately the basis of what we do. It is possible that some aspects of physical and mathematical research could be conducted on a desert island from first principles but there are few imaginable examples and they have, in fact, never occurred. Music and art can spring spontaneously, even calculating prodigies exist, but this is not the same as what we do. All of us have passed through a formative period in which we absorbed the cumulative results of the past centuries, the models, theorems, and empirical results of the preceding generations. Some of these have names attached, not always assigned properly by history but indicating, in some sense, the recognition that individuals have done, thought, or discovered something. Others have become so commonplace and essential that we no longer even note the authors, they may now seem obvious to us, but they were still human products.

It is in this light that the act of expropriating someone else's ideas, results, or words – plagiarism – is an offense, to those parties affected and to the scientific community as a whole.[28]

As I said above, the results of a scientific paper are common property but not its authorship. How a result is arrived at, understood, and explained can be as important as the result itself. Alone, a fact is nothing. Set properly in a broader picture it becomes a resource. If something cannot be understood by others it is worthless. We communicate in structures, words and equations and graphics, the results of our studies and these are particular to the source.

The concept of authorship is not just for "credit", a token to be used in job and grant applications, but that also plays a role. We live in a world that also judges what we're doing for the purpose of awarding (never forget the various nuances of that word) positions, observing time, support. Every time you apply for time at an observatory, beam time at a large national or international facility, a fellowship, or a position, a research grant, or even for a PhD program, individual contributions are the basis of the judgments. They are also the basis for choosing those who will be the judges. True, this is an idealization of a process but there is an idealistic side to the "scientific society", it *must* assume that those who are involved in all levels are fair and honest. For the cheater, to expropriate something gets ever easier. Previous generations had to work a little harder to be successful plagiarists. Before there were databases of published articles, or electronic reproduction, it was necessary to go to a library, or at least open a journal or book, and manually copy the material. The act required such effort that only the dedicated were willing to lift something. With a smaller community and fewer publications, it might even have been harder to get away with it. It was a deliberate, egotistic, conscious act. Technology allows anyone to cut and paste almost anything. This, along with the "democracy" of the web, seems to blur lines between the work of a colleague and yours. If something was said better in a published paper, why not simply use that yourself to explain something or describe something. If a figure serves your purpose, why not simply show it in a talk? Nobody will care because it is not the central point of what *you* are doing or saying, it is background.

But it is not. Fitting a curve to others data without citing their source is plagiarism. Changing the data to fit the model, or select those that fit and discard the rest, is fraud. And the opposite is just as true: showing a result, any result, without proper attribution and explanation of the criteria for the comparison is simply wrong and inadmissible.

---

[28] As an illustration of the point, as this chapter was being completed the German defense minister, Karl-Theodor zu Guttenberg, was forced to resign and renounce his PhD from the University of Bayreuth as a result of plagiarizing passages from other scholars that produced a protest by the academic community against treating this as a minor episode. As quoted in the NY Times (2011 Mar. 1, reported by Judy Dempsey)", ... last weekend more than 20,000 scholars from Germany and other parts of Europe sent an open letter to the Chancellery saying that Mrs. Merkel's continuing support of Mr. Guttenberg was a "mockery" of all those who "contribute to scientific advancement in an honest manner". "If the protection of ideas is no longer an important value in our society, then we are gambling away our future", the statement said. "We do not expect gratitude for our scientific work, but we do demand respect. The scientific community is suffering as a result of the treatment of the Guttenberg case as a trivial offense. As is Germany's credibility". You see, there can be very serious consequences.

One of the roles of the referee-editorial process is to help guard against such abuses, committed consciously or unconsciously. The ethics of authors is taken for granted, it is not the norm to assume that a paper is fraudulent or is misrepresenting the work (*e.g.* LaFollett 1996).

There is one additional form of plagiarism that, while not seeming serious, is unethical, the direct use of the text of the referee report. The referee may make suggestions on specific wording. In such cases, a properly phrased acknowledgment is sufficient to indicate that the authors have received direct "editorial" help from the reviewer. In other cases, however, the text of the report may contain a critique that is *not* editorial but intended as a critique to be addressed by the authors with textual revisions and further explanations (in the cover letter that accompanies the revised manuscript). This is *not* to be simply copied into the text of the revision, yet there are instances of such appropriation. This is laziness on the authors' part but it also does not address the critique, but merely acquiesces to the reviewer in the hope that the change will make the paper acceptable. In fact, it should do just the opposite since it shows that the authors are doing "whatever is necessary to satisfy the referee". To an editor this is absurd. It shows an ignorance of the point of the refereeing process and may even produce unresolved logical contradictions in the paper.

A recent development is the employment by a number of journals of automated commercial software to check for plagiarism. Functionally similar to search engines, they provide a "matching statistic" that signals potential problem cases. This *never* replaces the human evaluation but it is efficient and usually reliable as a first step, often being used because the editor notices something even before the paper is sent to a referee.[29]

4.8   Ghost and "honorary" authorship

It has been said that "on election day, the dead rise" and this also happens with journals. The phenomenon of ghost, or *honorary*, authorship is yet another strain on the ethical cords of scholarship. The stories are countless and, some, extremely sad.[30] The honorary authorship may lengthen a publication list but it is dishonest, both for those included and those who include them. It is sometimes done believing that it will ease the review process by seemingly giving a "stamp of approval" by senior and/or well known scientists to the work of a junior or less well known colleague. It may, however, have just the opposite effect. Should any of the "ghosts" discover their names in an author list of a paper they have not previously seen, *e.g.* after being posted on astro-ph, they may write directly to the journal, initiating an investigation and possibly leading to sanctions against the offenders. Some journals require an explicit statement of responsibility – who did what in the research and the writing (*e.g. Nature*), while others take this as implicit – as well as "conflict of interest" statements. In very large collaborations[31], there are usually

---

[29]This is well described in an editorial and several opinion pieces in *Nature* (2010, *Nature*, 466, 159 Jul. 8, see also Titus & Bosch (2010) and Koocher & Kieth-Spiegel (2010).

[30]One, however, is actually funny. The author list of Alpher, Bethe, and Gamow's paper on cosmic nucleosynthesis was the product of Gamow's peculiar sense of humor but Bethe was actually aware of the work and agreed to the $\alpha\beta\gamma$ ordering.

[31]This is a growing feature of astronomical research that has been commonplace in, for instance, particle physics for decades: the Virgo and LIGO collaborations for gravitational wave observations numbers in the

internal agreements that require participants to be actively involved in the work in some way during a period of time. It is a universal editorial "given" that the paper has been reviewed, and approved, by all concerned. There have been cases, fortunately not in astronomy, when a paper has been shown to contain fabricated results, in those cases embarrassed ghost co-authors have been held responsible along with the offenders, often doubly so because they are forced to use the explanation that they did not even know what was in the paper and, thus, cannot be held liable for the falsifications. Think about what this says about the scientist making such a defense (noting, mainly from very high-impact journals, some of the recent scandals in material sciences, biomedical sciences, and famous cases in paleontology; it is enough to recall the case of recent bubble "cold fusion").

4.9   The role of acknowledgements

In light of the last section, let us finally turn to the last part of a published paper. The role of acknowledgments is to make public the indebtedness of the authors to those whose counsel and help does not merit authorship but who, nonetheless, contributed in some way to the final paper. This is no mere social token. It makes clear the role of those who have been involved in the background. This includes the referee, even anonymous (it is not ridiculous to thank an anonymous reviewer, the referee and editor know who it is and it makes clear to the rest of the community that the refereeing process has been effective). It is particularly uncivil and uncollegial to *not* thank a referee when the review has been open. This section also provides the antidote to the ghost author attribution, an appropriate acknowledgment suffices.

5   Concluding remarks

It is clear from the history of science that the publication process, based on external peer review and scientific editorial oversight, is neither pure nor perfect. But it has evolved and can be improved. It is as unending an experiment as science itself.

My sincere thanks to Claude Bertout and Helmut Abt for their wise guidance and for innumerable, golden discussions and exchanges over the years. The Schools in 2008 and 2009 have been, for me, the experience of a lifetime and I thank all of the students for their enthusiasm, probing questions and critiques, and for their precious gift of a vision of the future of the field through their eyes. Thanks to Thierry Forville, Pierre Henri, Chris Sterken, and Joli Adams for their comments on this chapter and for our discussions during the School and for their wonderful friendship, and to my fellow editors at A&A in this amazing enterprise for their scientific and editorial collaboration. I thank Jason Aufdenberg and Pierre Henri for their detailed critiques of the text, and Daniele Galli, Enore Guadagnini, Jordi José, Rob Kennicutt, Ted LaRosa, Leon Lucy, John Mariska, Georges Meynet, Jan Palous, Francsco Pegoraro, Pier Giorgio Prada Moroni, Paolo Rossi, Luigi Stella, Ethan Vishniac,

---

hundreds, similarly for the Cherenkov telescopes MAGIC and H.E.S.S., large space projects such as *Fermi*, and surveys like the Sloan. Of course not everyone in groups of this size contributed at every stage, but there are internal refereeing groups, sites on which the preprints are posted as drafts for internal comments, and a host of other mechanisms for insuring that those involved take appropriate responsibility for the contents of the papers issuing from the projects. Often lead authors are identified as the contact points so they are visible even when an invariant order of authorship has been worked out by negotiations within the collaborations.

and Glenn Wahlgren for countless discussions over the last decade on these and related topics, I also want to thank the thousands of referees and authors with whom I've had the pleasure of corresponding over the years for many enlightening offline discussions.

Dedication

This chapter is dedicated to Cody (Three Rivers, Michigan, 14 May 1994 – Blankenberge, 19 May 2008).

References

Abt, H.A. (ed.), 1995, American Astronomical Society Centennial Issue of the Astrophysical Journal (Chicago: Univ. of Chicago Press)

Alberts, B., Hanson, B., & Kelner, K.L., 2008, Science, 321, 15: "Reviewing peer review (editorial)"

Aller, L.H., 1999, ApJ, 525, 265: "Menzel's 'Physical Processes in Gaseous Nebulae'", commentary in AAS Centennial Volume

Anderson, R., 1993, Notes Rec. Roy. Soc. London, 47, 243: "The referees' assessment of Faraday's Electromagnetic Induction Paper of 1834"

Batchelor, G.K., 1981, J. Fluid Mech., 106, 1: "Preoccupations of a journal editor", a must read for anyone wanting a deep reflective assessment of scientific publishing

Brown, C., 2004, Scientometrics, 60, 25: "The *Mathew Effect* of the *Annual Reviews* an the flow of scientific communication through the World Wide Web"

Eggen, O., 1993, ARA&A, 31, 1: "Notes from a Life in the Dark" (this is where Eggen lets it all out at the referees and editors)

Evans, J.A., 2008, Science, 321, 395: "Narrowing of science scholarship". This is one of the most important studies of citations and the evolution of the "general culture" of the literature, and one that has received considerable attention (by a historian of ancient astronomy, so one of wide culture)

Giles, J., 2006, Nature: on checking of results, "The trouble with replication"

Hartman, P., 1994, A Memoir on the Physical Review: A history of the first hundred years (NY: AIP Press)

Hoffmann, D., 2005, Ann. der Physik, 17, 273: " '... you can't say to anyone to their face: your paper is rubbish.' Max Planck as Editor of the Annalen der Physik"

Jackson, D., & Launder, B., 2007, Ann. Rev. Fluid Mech., 39, 19: "Osborne Reynolds and the publication of his papers on turbulent flow"

Kennerfick, D., 2005, Phys. Today, (Sep.), 34: "Einstein *versus* the Physical Review": this is a rare look into the archives at the process behind refereeing

Koocher, G., & Kieth-Spiegel, P., 2010, Nature, 466, 438: "Peers nip misconduct in the bud"

LaFollette, M.C., 1996, Stealing into Print: Fraud, Plagiarism, and Misconduct in Scientific Publishing (Berkeley: Univ. of California Press)

McClelland III, J.E., 2003, Trans. Amer. Phil. Soc., 93(2), 1: "Specialist Control: The Publications Committee of the Acadèmie Royale Des Sciences (Paris) 1700-1793": The most complete study to date on the origin of the refereeing system in the 18th century

Meadows, A.J., 2008, Science and Controversy: A Biography of Sir Norman Lockyer, Founder Editor of Nature, 2nd Ed. (London: Macmillan)

Merton, R.K., 1968, Science, 159, 56: "The Matthew Effect"

Parker, E.N., 1997, EOS, 78, 391: a personal account, "The martial art of scientific publication"

Pyenson, L., 2005, Ann. der Physik, 17, 176: " Physical sense in relativity: Max Planck edits the Annalen der Physik, 1906–1918"

Reingold, N., & Reingold, I.H. (eds.), 1981, Science in America: A Documentary History, 1900-1939 (Chicago: University of Chicago Press)

Rose, M. 1993, Authors and Owners: The Invention of Copyright (Cambridge, MA: Harvard Univ. Press)

Schultz, D.M., 2010, Scientometrics, 86, 251: "Rejection rates for multiple-part manuscripts"

Titus, S., & Bosch, X. 2010, Nature, 466, 436: "Tie funding to research integrity"

Wali, K.C., 1991, Chandra: A Biography of S. Chandrasekhar (Chicago: Univ. of Chicago Press)

Wright, H., 1966. Explorer of the Universe, A biography of George Ellery Hale (NY: Dutton)

Zucker, R.S., 2008, Science, 319, 32: "A peer review how to"

Some additional readings:

Abt, H.A., 2000, unpublished: a personal note, this set of comments on refereeing was prepared during Abt's term as editor of the *ApJ* but was, alas, unpublished. It is, however, availble from its author

Berendzen, R., 1974, Physics Today, 27: 33-39: "Origins of the American Astronomical Society"

Burrell, Q.L., 2005, Scientometrics, 65, 381: This is a particularly interesting paper, "Are sleeping beauties to be expected?" The phenomenon is a paper that, after years of neglect, suddenly becomes "hot". Perhaps the best example I can cite is the famous EPR (Einstein-Podolsky-Rosen) paper on the completeness of quantum mechanics. For decades it was known, and discussed, almost exclusively in the philosophical and historical literature, along with Bell;s theorem. Then along came quantum computing and suddenly these became required reading at the leading edge of theoretical physics

Chilton, S., 1999, Academe (Nov.) 54: this paper is one of the sleeping beauties, intended for a general academic audience (this is the magazine of the American Association of University Professors (AAUP), the group that defined tenure in the North American university): "The Good Reviewer"

Cole, J.R., & Cole, S., 1972, Science, 178, 368: another example of the sociological view of the referee system, following the same path as Merton's studies, "the Ortega effect"

Cole, S., Cole, J.R., & Simon, G.A., 1981, Science, 214, 881: another of the social studies similar to Merton and the Matthew Effect, "Chance and Concensus in Peer Review"

de Solla Price, D., 1975, Science Since Babylon: Enlarged Edition (New Haven: Yale Univ. Press) a real classic, one of the first historical-social studies of the growth of science by a leading historian of physical science, see chapter 8: "Diseases of science"

Garfield, E., 1986, Current Contents, nr. 32, 3: one of the classics, frequently cited, by the founder of Science Citation Index, "refereeing and peer review"

Goudsmit, S.A., 1963, Phys. Rev. Lett., 10, 41: "The future of physics publications: a proposal", an editorial at the start of the literature explosion in the 1960s

Greene, M., 2007, Nature, 450, 1165: as the number of authors multiples, and projects become ever more complicated (in fields that previously had been immune to the "big science" team research structure), this paper offers some interesting insights: "the demise of the lone author"

Holmes, F.L., 1987, Isis, 78, 220: a general historical account, "scientific writing and scientific discovery"

Judson, H.F., 1994, JAMA, 272, 92: one I recommend for its view of the process in perhaps the most sensitive field for refereeing, medicine: "structural transformations of sciences and the end of peer review"

Phelan, T.J., 1999, Scientometrics, 45, 117: a guide for the perplexed in bibliometrics, "A compendium of issues for citation analysis"

Skilton, P.F., 2006, Scientometrics, 68, 73: A recent study of the evolution of citations as they propagate through the literature, "A comparison of communal practice: assessing the effect of taken-for-granted-ness on citation practice in scientific communities"

Suppe, F., 1998, Phil. Science, 65, 381: "The Structure of a Scientific Paper" (see also comments by Alan Franklin and Colin Howson); also Lipton, P. ibib, 406: "The best explanation of a scientific paper"

Weller, A.C., 2000, J. Amer. Soc. Inform. Sci., 51, 1328: one of the early reviews, "Editorial peer review for electronic journals: current issues and emerging models"; see also Brown, C. 2001, J. Amer. Soc. Inform. Sci, 52, 187: an example of the process of expansion of the arxiv viewed from outside, "e-volution of preprints for physics and astronomers"

Scientific Writing for Young Astronomers
C. Sterken (ed)
EAS Publications Series, **49** (2011) 63–70
www.eas-journal.org

THE ROLE OF THE PUBLISHER

A. Henri[1]

Abstract. This paper describes the various facets of the publishing activity. The task of the publisher is to make every effort to promote the work of researchers who have submitted their article for publication. This requires the set up and maintenance of an effective management interface, timely production, delivery of quality items, a high-performance web interface, sustainable data archiving, indexing in large databases, and round-the-clock maintenance of data access. It is a daily work that envolves a dedicated professionals team for the timely production of *Astronomy & Astrophysics*.

1 Introduction

Managing a journal, and especially a major astronomy journal like *Astronomy & Astrophysics*, requires time, energy, and professionalism. Most people think that typesetting is the prominent part of the publishing enterprise; because the authors use LaTeX macros, the publisher no longer seems really necessary. Although it is true that, compared to twenty years ago, the publisher is less involved with the layout, a lot more time – hence expenditure – is now required for the editorial management, the online publication, the archiving and indexing of the data, etc. EDP Sciences (hereafter EDPS) has been publishing physics journals and series of books since 1920, and thus has acquired sound experience in the production and publication of scientific journals.

EDPS is a subsidiary of learned societies and, as such, its prime aim is to participate in the dissemination of scientific information, rather than the mere pursuit of profit. For the research community as a whole, community-owned journals represent the best vehicle for distributing the scientific findings, are the optimal media for promoting the scientist's work, and last but not least, act as significant catalysts for the individual research careers. The publisher is the partner of the scientific community, and the partner's role is to work in collaboration with the editorial team of scientists on the following main points:

1. improve the papers during the refereeing process with the technical services of the publisher,

---

[1] EDP Sciences, 17 avenue du Hoggar, BP. 112, 91944 Les Ulis Cedex A, France

2. be a reliable source of information (only validated versions are published), and

3. promote the work and career of researchers.

Our mission is therefore to provide the A&A scientific community with all necessary tools for enhancing the visibility of their work at the lowest possible cost for the subscriber.

The following sections describe the different parts of a publisher's job: from the management tools to online publication, through production, marketing and web development.

2   Management tools: from submission to acceptance

To produce a journal such as A&A, which receives around 2500 submissions per year, the editorial office needs a high-performance tool for managing all documents. Indeed, the editorial office of a journal is the very first contact that scientists have when they submit their manuscript. In this respect, it must guarantee efficient service and satisfaction for authors. An efficient software tool is needed for submitting manuscripts and following the peer-review and handling of articles by each scientific editor. The A&A editorial office uses the *Manuscript Management System* (MMS), a manuscript management software package developed by EDPS, which was specially designed to help editors, authors, and editorial staff manage in real time the manuscripts under evaluation. The maintenance and development costs for this tool involves two positions (software engineer and product manager) at EDPS.

3   Production: from acceptance to publication

A high-quality scientific journal must feature efficient, high-quality production and a fast publication time. The production involves different steps that are summarized in Figure 1. The production process is more and more managed like an industrial process, with the aim of reducing delays and costs.

3.1   The first step: copy editing

After acceptance of a manuscript, the authors send the files of the manuscript to a dedicated e-mail address or via the MMS. A team of production editors takes care of the manuscripts with regard to

- check of the spelling, typographic mistakes, etc.;

- preparation of the references for use by CrossRef[1] in the online version;

- consistency check of graphics; and

---

[1]CrossRef's mandate is to be the citation-linking backbone of all scholarly information in electronic form, see http://www.crossref.org/

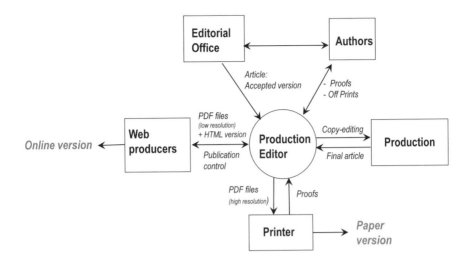

Fig. 1. Production workflow from acceptance to publication.

- content consistency: make clear distinctions between physical variables, mathematical symbols, units of measurements, abbreviations, chemical formulas, etc.

We mentioned earlier that compared to two decades ago, authors now do a lot of type-setting labor when preparing their manuscript by using, for example, the LaTeX macros provided by the publisher. Nevertheless, they are not fully aware that, quite often, small errors remain: authors are not expert typesetters and, because they concentrate more on the content than on the form, they make mistakes in the placement of figures, they incorrectly quote page and volume numbers in the bibliography, etc. Figure 2 (which should be viewed in color) shows an example of a copy-edited bibliography page in A&A. It clearly demonstrates that numerous corrections have to be made in order that links towards other journals work.

3.2 The second step: prepress

Once the manuscript has been proofread, the EDPS typesetter brings in the corrections required by the production editor and applies the layout with the A&A LaTeX macro. For the figures, the typesetter verifies consistency of graphics (*e.g.*, by cleaning the figures, correcting the textual components in graphs, adjusting thickness of lines, checking the colors – *i.e.*, *RGB* or *CMYK* depending whether the figure is for printing or for the online version, incorporating all fonts needed to avoid any problems during the final printing process, fixing the resolution, etc.). Once this process is finished, the manuscript PDF file is returned to the production editors. They then check that everything is correct and send a proof to the author for final correction and approval. Once the corrections by the author have been applied and after a final check of the paper's contents by the editorial office, the

Fig. 2. An example of copy-editing of the bibliographic references in A&A.

production editor inserts a *Digital Object Identifier* (DOI) number[2], and the article and volume number. The article is then ready for publication.

[2] A DOI number is attributed to each paper (http://dx.doi.org), ensuring the longevity of the article's URL. Together with CrossRef, the DOI identifier also allows for including a direct link to the publisher's website of the referenced articles in the online version of the bibliographic references.

3.3 The third step: publication of the online version

The files are then processed for the web version: preparation of metadata, full-text and references in XML (the XML format allows for the creation of the HTML full-text web version, and is currently the format needed for data sustainability). In the newly introduced system where articles rather than pages are numbered, the journal volume is built up in a progressive manner and is brought to a close when an adequate number of pages have been published.

3.4 The fourth step: the paper version – printing and dispatching

When enough articles and pages have been produced to create a volume, the production editor prepares it (by organizing contents, cover, etc.), and sends it by ftp to the printing house. Production editors are in permanent contact with the print manager to solve any technical problems that may arise, and each stage of the printing process is checked.

4 Web platform

Maximizing the distribution yields more visibility to articles and improves the impact factor of the journal, thereby promoting the results presented in the articles, their authors, and the institutions that employ them. What is therefore required is a state-of-the-art, efficient web platform offering numerous services and advanced features for distributing the online edition of the journal. The A&A platform includes several tools, as follows.

CrossRef indexing and hyperlinks in the bibliographic references. There are approximately 2550 participants in the CrossRef system (publishers, research companies and institutions, libraries), representing more than 20 000 journals. The system enables the creation of millions of links between the different journals. EDPS also provides direct links to referenced articles in other databases such as NASA/ADS and MathSciNet. Readers appreciate these features, because they allow them to easily navigate between different articles in a single domain issued by different publishers.

Online publication of the HTML and PDF versions, electronic-only material, and archiving. As we described above, the publisher should provide a full-text HTML version of each paper, and also allow for electronic-only material designed to provide supplementary information that is either too voluminous for printing, or designed specifically for the Web, such as large tables, appendices, programs, images and movies, etc. Another mission of the publisher is to archive the contents and to ensure permanent access to the journal, by

- hosting the data for an indefinite length of time;
- providing XML archives and PDF files;
- retro-converting data from time to time in order to maintain the accessibility to the contents;

- delivering abstracts to databases such as ADS-NASA, ISI, and Scopus;

- linking to data available at the CDS, and

- providing a platform that gives rapid access, even when traffic levels are high, with site maintenance ensured 24 hours out of 24, 365 days a year.

Specialized services – reader information.   The publisher's platform should offer a wide range of services for the reader:

- Email alert and RSS feeds.

- "Citing article": this returns a list of articles that cite the selected article. The citing articles come from the EDPS Electronic Journal database, as well as from other publishers participating in the CrossRef Forward Linking System.

- "Articles from the same author": this provides a list of papers written by the same author in the EDPS database.

- "Related articles": this furnishes a hyperlink to articles with a common author or common keywords in the EDPS database.

- A website fully indexed by Google, and participating in Google Scholar[3].

- Database collaboration, *i.e.*,

    - CDS: EDPS, in collaboration with CDS, improves the author-supplied links pointing to astronomical objects in the CDS databases and virtual observatory tools.

    - ADS-NASA: EDPS supplies the complete archives of A&A Supplement, returns XML code conforming to the ADS technical specifications for each new issue of the journal.

- Dexter application: ADS has developed the applet Dexter, which has been adapted by EDPS to work with A&A articles. This software provides the capability of exporting data from the figures.

- Search engines:

    - CrossRef Search: this pilot system, put in place by the CrossRef consortium, is a free search engine that facilitates navigation in scientific journals by specifically indexing scientific articles. The search engine uses Google technology and DOIs.

---

[3]Google Scholar is the academic version of Google.

– Intuition search engine: this advanced search engine provides multi-criteria searching: author, title, keywords, addresses, publication date and complete text. Searches can be made in a particular journal, in a particular scientific domain (physics journals, for example), or in all the journals published by EDPS.

• OAI Server (Open Access Initiative): EDPS has put in place an Open Access Initiative depot, which will make available articles metadata in a standard format (DublinCore) at no charge, thus enabling the interoperability of different OAI archives. Furthermore, the Publisher must permanently analyze and evaluate new emerging technologies so as to offer more and improved services to the journals and their readers.

The maintenance of the web platform including web hosting, software management, and software development keeps a team of several people busy at EDPS.

5   Communication and marketing

The marketing and communication department's aim is to maximize the journal's visibility to researchers, authors and readers. The more people become aware of what happens in the journal (publication of high-quality papers, special issues, etc.), the more people will read and cite its articles, thus bringing to the authors the visibility they need. A team of marketing assistants must

• print promotional material (a leaflet for Astronomy & Astrophysics) and distribute it as widely as possible,

• carry out promotional items that will disseminate additional support materials for authors and readers,

• send information and announcements to readers (on a new section, a highlighted paper, etc.),

• represent the journal at international scientific meetings (AAS meetings, IAU General Assemblies, SWYA school, etc.),

• promote selected articles: some outstanding articles merit being brought to the forefront by the editors. For these articles, EDPS works in partnership with the editorial office, to

– make them freely available,

– highlight them by a specific category ("Press releases" or "Highlighted Papers") that is directly accessible from the homepage of the journal,

– add an accompanying text to explain the originality of the article.

6 Conclusion

The publishing business has changed significantly in recent years. Expertise has shifted to activities that are much wider than layout and dissemination of the printed edition of an article. Advanced technological tools are required (article management tools, an up-to-date publishing platform) with specialized technical personnel. A dedicated team of several professionals works at EDPS to give A&A the highest impact.

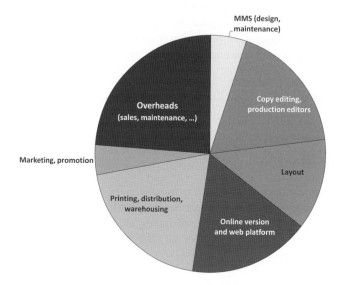

Fig. 3. Sample distribution of publisher cost factors (*i.e.*, non-editorial publishing expenditures).

The list below summarizes the cost factors on the publisher's end, and Figure 3 illustrates the relative shares of each item in the total operational cost.

1. development and maintenance of the Manuscript Management System;

2. copy editing as a consistency check of all aspects of the manuscript (figures, references, etc.);

3. prepress work (layout);

4. publication of the online edition (PDF, HTML, online material, ...), electronic-only figures and tables, etc.;

5. printing and dispatching;

6. archival services, and

7. specialized services (dedicated web site with many functionalities, marketing, etc.).

I thank Chris Sterken, the Editor of this volume, for proofreading the paper.

Scientific Writing for Young Astronomers
C. Sterken (ed)
EAS Publications Series, **49** *(2011) 71–83*
www.eas-journal.org

LANGUAGE EDITING AT *ASTRONOMY & ASTROPHYSICS*

J. Adams[1]

Abstract. In 2002, the A&A Board of Directors voted that all articles must be written in English and decided to improve the overall quality of the language in the articles with the help of a team of language editors. This article reviews the general advantages of editing the English expression and describes both the aims of this effort and its place in the full publication process. This is followed by the Guide to language editing that has been available on the Journal's website for several years now.

1 Background

The very first volume of *Astronomy and Astrophysics* contained only three articles in French compared to the 60 in English, and none were published in the other European languages representing the consortium of the countries that founded the Journal in 1968, although there were occasional submissions later in French and German. At first the authors were responsible for writing understandable English, so a paper's English was looked at only if it was full of errors. At times, the Editors suggested changes, as would the referees; but when it was unreadable, it would either be sent back to the authors asking that a colleague revise the language or else sent to someone with excellent English known to the Editors to revise it because the authors had no resource in their institute for the task.

This early arrangement worked because so many European authors had already been writing in English. It is also understandable, since this new journal had already achieved a lot and was still developing its policies and procedures, so that improving the English expression any further had to await other advances, in particular adding sponsors and then moving the editorial operations to a single site, thereby freeing up some of the operational expenses for more than one full-time language editor.

Most papers in A&A have always been written by non-native English speakers, and they have also been read by non-native English speakers over the world, so in the end the main goal has always been to seek clear and correct English when the A&A editors, referees, and now language editors make suggestions for changes. Like all other changes at the Journal, it hopes to provide its authors with the best vehicle for presenting their

---

[1] A&A editorial office, Observatoire de Paris, 61 Av. de l'Observatoire, 75014 Paris, France

©The authors, published by EAS, EDP Sciences, 2011
DOI: 10.1051/eas/1149005

research and ideas in an international forum. Unclear or inefficient English only stands in the way of this goal.

This article covers the background to the role of English in the Journal, which of course includes why English is so important in international science (Sect. 2). Section 3 tries to explain why the quality of the expression, in particular, is important, so it goes into some of the language problems worked on in each paper. Section 4 describes the process of language editing within the Journal itself in more practical terms and is followed by some suggestions for other resources that authors can use before and after submitting papers to the Journal. After a short concluding statement, the *"Guide to English editing at A&A"*, already available on the A&A website, is printed in this volume.

2   Why move to only English at A&A

In 2001, the A&A Board made official an earlier policy to publish only in English at A&A. This decision was in line with the situation in other scientific journals and with all the other changes made at the Journal over the years to improve its effectiveness as a forum for European science in the world. The attempt to improve the quality of the English began at about the same time, perhaps, and is one aspect of the more general improvements and only one of the many reasons that the Journal's impact has been rising.

Few papers written in other languages had been accepted for many years, but not because they were written in French or German. In fact, few were being submitted in any other language than English to start with, because authors were aware that it would restrict a paper's impact in the scientific community, no matter how good the science. The Editors' report to the A&A Board in May 2002 shows this awareness:

> A few authors from France kept submitting papers in their national language. This causes two problems: first, papers in French have a limited readership in the international community and thus contribute to lowering the impact factor of A&A, and second they can only be sent to referees with a good knowledge of French, which
>
> a) dramatically constrains the choice of referee,
>
> b) does not always insure that the most competent referee is used for a given paper, and
>
> c) creates a *de facto* inequality between A&A authors.

These problems led the Board to decide to modify the Instructions to Authors to make it mandatory for all submissions to be written in English. Until the first hiring of a part-time language editor in 2000, only the most incoherent papers were sent to someone to have just the language corrected whenever needed, so that all the other papers were corrected with the help of either a friend or the referee, or not at all.

2.1 *English as a* lingua franca

English is the global language in many domains, not just science. It is the *lingua franca* at meetings where participants come from several cultures and speak many languages, much like Latin was in the West in the first millennium.

In my opinion, the significant reason English is used is not that, for example, the British, US North Americans, or Australians speak it as a first language. If this were the case, then we should all be using Spanish instead because it has as many native speakers as English. Even better would be Chinese, not English, because there are more first-language speakers of Chinese.

The deciding factor is that English has become the main second language of all the other cultures in those areas where global communication counts, and it has become so even more since the fall of the Iron Curtain. As a language, English is not any better able to adapt to this role than any other language, at least not linguistically better: it just happened that way in this millennium. Wikipedia has several articles summarizing this role of English, if the reader is interested in exploring it further.[1] That hegemony as a global language may change in future millennia, who knows?

In line with this use of English, A&A is not just a vehicle for publishing scientific articles. It also reflects the growing community of astronomers and astrophysicists from over 60 countries for whom publications in English are the main medium of communication. It takes that role as a forum seriously by setting up ways for less advantaged scientific groups to publish their research, and the language editing service has always played a part in this effort. Between the reader survey and the editorials about policies, it also believes in the benefits of communicating its procedures, as the reader will discover from the various materials intended to help the author through the language editing process, found online and at the end of this article, and by our availability to answer questions about why certain suggestions are made for the English expression.

2.2 *Editing for clear and correct English expression*

Besides the main reason cited above for using English, A&A soon began to concentrate on improving the overall clarity of its published articles by introducing language editing to make certain a good level of English is maintained. There are several reasons for this, beginning with the obvious desire to be read and understood by the most possible readers in the field. I list some others here.

- Since language is edited in almost all professional publications in all languages, A&A is only showing its professionalism, thereby providing the best possible forum for the research.

- Most of the Journal's authors are not native English speakers, so this move by the Journal is an attempt to keep authors from being unfairly restricted by the need to publish in another language. It is by no means an attempt to add just another step in the already complicated one of publishing.

---

[1] See http://en.wikipedia.org/wiki/English_language#English_as_a_global_language

- The job of the language editors is to allow the science to appear as it is intended. If the language is awkward or garbled, the science may be interpreted as unclear, too.

The language editors have been hired to help the Journal's articles show more coherence in the language and form of all papers, and this remains the goal, even if there are a few differences in how each separate paper is treated. To achieve this, the language editors rely on grammar rules and the precepts of good style – the plain sort – they were all taught and trained in from the start. And to a certain degree, they must depend on a "feel for what sounds natural," even if not able to define a specific rule for the suggestion.

3   What is edited: seeking quality

The language editors do not try to change the science in any way. That is the job of the referee and scientific editor. Instead they look at the language and some details of presentation, such as the figure or table captions, paragraphing, and the abstracts. The goals of A&A English language editing are to correct

- grammar and vocabulary, including inconsistent spelling,

- ambiguous sentences,

- colloquial or familiar language,

- awkward expression, so it is easy to read with smooth, flowing sentences that are not too long or complex.

In other words, the language should be correct, consistent, unambiguous, and formal, and it should also carry the content, so that the article reads smoothly and clearly. The language should flow as the reader continues through the text, with nothing getting in the way of understanding what is being said, any more than will occur for those whose English is not yet a high level.

Review by a trained, but non-specialist reader allows areas of text to be detected that are not very clear or that may be confusing to some readers, including to new scientists or those not in the author's own specialty. To this end, the style of a scientific article should allow the reader to pay full attention to the scientific content and not to how complicated, elegant, or even awkwardly it is written.

Another factor is the attempt to create a professional context for presenting the author's work, which is also taken into account in our suggestions. Besides the goal of a serious and formal style, they also aim for consistent spelling and word use, which is another measure of a publication's level. The second section of the *"Guide to English editing at A&A"* states this principle and goes into detail on the main inconsistency to correct: mixing British and US American spellings. We do not require one or the other at A&A, but expect it be the same throughout each single article. It may be a detail as far as clarity is concerned, but is one that is considered an important detail in the publishing world.

3.1 Why simple, as well as correct, English

We read the text to assess its grammar, syntax, and clarity. We intervene first to remove grammatical errors and inconsistencies in spelling or word choice, second to resolve any ambiguities of expression, and third to smooth out or simplify the style. Editing for a simple and correct English basically means that overly poetic language with its metaphors and elegant phrasing is discouraged, unless the writer has full control over it and a good reason to use it. Every reader should be able to follow the ideas without difficulty. In this case, "simple" does not mean "simplistic" or "simplified", but rather:

- use clear and plain rather than elaborate,

- use the main dictionary meanings of words not their literary or extended senses,

- avoid metaphor, jargon, or long convoluted sentences.

This is the opposite of what it may seem as your words are rearranged and alternative phrasing suggested. This is not the style of, say, Shakespeare in his poems and plays. Since he was writing before any grammar or spelling books had been written for English, unlike Latin or Greek, the language editors would now be forced to correct his originals much more than any science article seen nowadays, but they would hardly try to reproduce his specialty, the exquisite poetry and rhetoric, as in

> Thus Conscience does make Cowards of us all,
> And thus the Native hew of Resolution
> Is sicklied o're, with the pale cast of Thought,
> And enterprises of great pith and moment,
> With this regard their Currants turne away,
> And loose the name of Action.[2]

But A&A would suggest he change to the literal expressions, not the poetic figures of speech, not to mention all the spelling, word choice, and capitalization problems this contains (*e.g.*, hew, currants, turne, loose), which no longer show up in modern editions.

Even then, the editors make certain that it will be understood by all readers, regardless of their linguistic background. Poetic language is by nature ambiguous with its multiple layers of meaning. This is the policy for all publications of scientific journals in the United States and the United Kingdom, if not elsewhere.

In the very useful *"Style Guide of the American Chemical Society"*, which is run in conjunction with Oxford University Press,[3] I find two quotations on this question:

> Long words and complicated sentences are not essential features of good scientific writing, although they are often thought to be so. The best writing in science, as elsewhere, is simple, clear, precise, and vigorous. Decide what you want to say and say it as simply, informatively, and directly as possible. (M. O'Connor, *Writing Successfully in Science*),

---

[2] *Hamlet* Act 3 scene 1, Folio I (1623).
[3] http://www.oup.com/us/samplechapters/0841234620/?view=usa

and

> In scientific writing, there is no room for and no need for ornamentation. The flowery literary embellishments, the metaphors, the similes, and the idiomatic expressions are very likely to cause confusion and should seldom be used in writing research papers. Science is simply too important to be communicated in anything other than words of certain meaning.

The second citation is from Robert Day in *Scientific English: A Guide for Scientists and Other Professionals*, and it goes on to say that "The meaning should be clear not only to peers of the author, but also to students just embarking on their careers, to scientists reading outside their narrow discipline, and especially to those readers (the majority of readers today) whose native language is other than English".

In this spirit, A&A also asks that the jargon and specialist phrases used among collaborators be avoided except when there is no other choice, because it may not be as familiar or clear to the other scientists reading the article; likewise, they are never as precise as the standard form, unless it is the only technical way it is expressed. If the language editors question the use of words that you feel are the most suitable, they are not challenging the science, but have instead seen a problem in the phrasing or effectiveness of its presentation. If a standard set of words exists in English for the same idea, then please use it instead of the specialist phrasing, even if it is present in most of your colleagues' papers, or try to vary between the two.

Likewise, there are some wordy expressions that show up in scientific and other specialized writing, so that untrained writers seem to feel they are more precise or more scientific than the standard form. Indeed, they are found in native English-speakers' papers, so technically correct, but it is perceived as bad style and unprofessional. When there are too many in the same section, it is either confusing or awkward.

In English, the main principle is that the more unnecessary words used, the less clarity in the message. Examples of some phrases include

> **be in agreement with, of the order of** (or **on, in**, depending on the dialect), **small number of, despite the fact that, show evidence of the presence of, owing to the fact that** (or any "the fact that"), **are known to be, at the present time, of great importance.**

Each of these should be replaced by their direct and shorter synonym as follows in the same order:

> **agree, about** (**roughly**, etc.), **few, although, proves the presence of** (or **evidence of** without **presence of**), **because, are, now, important.**

As you can see, even with the parentheses (similar character counts in both), the second list is much shorter than the first. This list was adapted from the much longer one on the Oxford and ACS page mentioned above. There are others that are seen in A&A submissions that are listed in our resources:

- the appendix to the October 2008 editorial on language editing,[4]

- "Some Frequent Corrections",[5] or

- Section 6.5 of the *"Guide to English editing"* on the site and reprinted as a chapter in this book.

That conciseness is an ideal in all or most English writing is supported by several experts and is one of the main things taught in composition classes in all English-speaking environments, so aiming toward that goal as much as possible is another consideration in the language editing, although clarity is still the main one. The most famous statement of this principle comes from William Strunk, Jr. in *The Elements of Style*:

> A sentence should contain no unnecessary words, a paragraph no unnecessary sentences, for the same reason that a drawing should have no unnecessary lines and a machine no unnecessary parts. This requires not that the writer make all his sentences short, or that he avoid all detail and treat his subjects only in outline, but that every word tell.

It has been shown over and over that using more words than needed leads to confusion so that the point in a sentence, paragraph, or article is lost. The *"Guide to English editing"* and *The Elements of Style* offer several other suggestions for writing more concisely and clearly, including examples of how to make your sentences more active or to express parallel ideas in the same structure.

3.2 Why formal English?

Asking for formal English stems from its being standard across all the national dialects of English, which is not the case for the informal versions of English. All standard formal versions of language aim at communicating clearly across differences, while familiar language like slang is meant to be specific to a single group or place, *i.e.*, to exclude others from the message. Informal or colloquial English creates a slightly different problem for both writers and readers than elaborate language, because it

- cannot be easily understood in all regions. There are too many variations that do not carry the same meaning between regions and dialects, even in the same language. This is not the case with the formal standard written language.

- is hard to use correctly or precisely. This is true for native speakers, so even more so for non-natives, and informal language changes too quickly for it to be used in archival documents, like a scientific journal should aim to be.

Once again, Wikipedia's writers summarize the situation very well:

---

[4] See http://www.aanda.org/index.php?option=article\&access=standard\&Itemid=129\&url=/articles/aa/full/2008/41/aa10787-08/aa10787-08.html

[5] At http://www.aanda.org/content/view/230/214/lang,en/

Formal written English is a version of the language that is almost universally agreed upon by educated English speakers around the world. It takes virtually the same form no matter what the local spoken dialect is. In spoken English, there are a vast number of differences between dialects, accents, and varieties of slang. In contrast, local variations in the formal written version of the language are quite limited...The differences in formal writing that occur in the various parts of the English-speaking world are so slight that many dozens of pages of formal English can be read without the reader coming across any clues as to the origin of the writer, far less any difficulties of comprehension.[6]

4   The role of language editing in the editorial process

Language editing is an integral part of the editorial process at A&A. When a paper is first submitted, the scientific editor may choose to send the paper back to the author requesting that the level of language be improved before sending the paper to an external referee. This is usually because the editor has determined that the paper contains enough scientific content to warrant review, but that the level of English hinders appropriate evaluation or that the paper will require undue language revision to reach the level sought at A&A.

This step acknowledges the difficulty of having to express oneself clearly in written English, regardless of one's first language, and the disadvantage relative to native English speakers of being required to publish in English. Having a manuscript clarified by English-language colleagues before submission can, therefore, speed up the publication process.

Before acceptance, the referee may give advice for improvements in the language along with the scientific presentation, or else recommend it for language editing at the Journal. Editing language is not the main function of the referee, but some do it as they are commenting on the article. Before language editing was made official, the referee played a larger role in this, but not all referees feel qualified to correct English, even when their level is higher than the authors' level in English. Instead, they are now only asked to make recommendations for whether revision of language is needed.

When the paper is being accepted, the scientific editor may make a recommendation that a paper be edited for the English and this is considered in the final judgment. A&A now has a single language editor who looks at every single paper for the quality of the language in the article. Besides scanning them to see whether the language needs revision, this language editor also looks closely at all abstracts for correctness, clarity, and conformity to the goal and style of an abstract. At this point the paper is then (i) sent on for language editing, if needed; (ii) sent on directly for publication; (iii) after the first page is corrected, sent back to the author for a final pass (sometimes with some suggested changes later in the article). (iv) In cases where there is a native English-speaking fellow author, she or he might be asked to look through it one last time again for minor corrections.

---

[6]See http://en.wikipedia.org/wiki/English_language

Currently, about 70% of the accepted regular papers are sent to a language editor, and the median time spent with the editor is 11 days. Most Letters to the Editor are seen by a language editor and corrections are done within one or two days.

4.1 How to react to suggestions

Corrected versions are returned to the authors for confirmation and corroboration. The author then should incorporate those changes into the LaTeX file of the paper and resubmit for publication. At this point, there are three choices.

- You agree with all the suggestions and resubmit to the office in Paris following the instructions in the email from the language editor. If there are no other problems, it will be sent on to the publisher directly. At this point it is copyedited by the publisher and the proofs sent to you for confirmation.

- You remain uncertain about some of the suggested changes and would like further explanations. Write an email with your questions to be sent to the language editor, who will answer you as soon as possible.

- You do not agree with a proposed change, because it changes your meaning, especially the scientific import. If a change was suggested, it means that the original has some problem in the expression, and the suggestion was the most logical with the words being used in the original.

If there are several such issues, it is helpful to have the preferred changes marked in bold face in a referee format version of the text uploaded on the A&A MMS website. The preferred corrections are almost always accepted provided no additional language errors are introduced. In matters of scientific content, the author's preference is always respected. It is also likely that the language editor will take the time to explain the suggested changes you are questioning, if your new version is not correct English. Some of the explanations can be found in the *"Guide to English editing"*, of course, and the original files were originally designed to be used at this stage.

5 Other aspects to look at

When reading a paper for the English expression, the language editors also take a look at the abstract, figure and table captions, equation punctuation, and the reference list to make certain they conform to the principles of the *Instructions to Authors*.

Typical problems that are dealt with in abstracts are

1. an incomplete abstract or one that is not in the expected order (the headings), or one where, for instance, the context is too long but *Aims*, which is obligatory, is too short.

2. an incomplete sentence when the headings are used, because the author follows *Aims* or *Method* as if it were a question, *i.e.*, the noun phrase that describes the aims or method. We expect each sentence to be a full sentence, such that the whole abstract would read coherently if the headings were removed.

3. references, including self-reference, in the abstract, which should be self-contained, that is, should only refer to the paper that follows, not to the background information expected in the introduction.

There are exceptions to this last rule (a direct dependence on another work, as when commenting on a specific result published by another group rather than simply working from their results, as all science does), but it should be avoided whenever possible. If another paper must be cited, then the full reference should be included, since abstracts are often published without their papers and reference lists. The editors will query every single reference in an abstract and ask the author to check with their scientific editor or the Chief Editor if there is a question of appropriate reference. If the present paper is a follow-up to an earlier paper in a series, the "Paper I in the series" is enough to use in the abstract, without the full reference.

The main problem found in the table captions is when authors try to include as much as possible about it: all the references, legend details, full description of column headings and of results. If an author is tempted to say "Note how the…" then it is discussing results that belong in the running text not the caption, which should refer exclusively to the legend symbols in the figure. There is a difference between table and figure captions because a table has a title and notes, while figures have short captions that explain the graphical information. There should be a minimum of repetition between text and caption, the exception being the brief description of the subject of the table or figure (in title form without many details) followed by the minimum of what is needed to follow the figure or table in a legend form. The Instructions to Authors give more specific instructions, but the only exception to these long captions are for figures that are published separately online, so that the full descriptions of columns and legends and references are needed, because the text is not directly available, such as it is when it is embedded within a full text.

Even though the equations are part of the scientific editing, it follows the syntax of the sentence that it must fit into, what is called embedding, so is also part of the language. They may never stand alone, but instead continue the preceding sentence as an example following a colon (":") or directly with no punctuation at all before it begins, even when separated. The sentence can continue after the equation if needed, but then no colon or semi-colon (";") should be used more than once, as in a normal sentence without any equations. Likewise, when the sentence ends with the equation, then a period (full stop) follows it directly. This is the house style, so it will also be checked by the copyeditors at the publishers.

The reference list is glanced through for potential problems, although a more thorough job is done at the publishers. If there are arXiv references without indicating what the paper's status is with the journal where it will be published, then it will be queried and a reminder goes to the author to make certain that this is included in the final version. There are few cases where A&A accepts work that has not been refereed, so these cases will need to be gone through with an editor.

5.1 Recommended resources when writing for A&A

The following resources are recommended for answering questions about suggestions for changes and why. The language editors have been working together to provide you with most of the guidelines we use when correcting your English expression. The *"Guide to English editing at A&A"* is for your use at any moment. It does not cover every single rule of English usage (*e.g.*, use of articles for languages that have none or few, because it cannot, but it discusses those that are found most across all the authors' linguistic backgrounds.

1. Author's guide on the A&A website, starting with the Instructions to Authors. This covers everything except language use. We refer authors to it when suggesting corrections to figure and table captions or to the reference lists.

2. The other pages for authors, including the "Language editing" pages and its attachments. The Language editing page summarizes the policy of the Journal. The attached files include a PDF file that is easy to download for future use: *"Guide to English editing at A&A"*. You can also go to the "Some frequent corrections", which is an HTML file in chart form showing how to correct some phrases correct a lot on A&A articles and what suggest as a better way to say it. The explanations for many of them are in the *"Guide"*, but not all, since many are vocabulary problems, such as which preposition goes with which noun or verb.

3. Your own bilingual dictionary but complemented by a monolingual dictionary if possible. You can always use the online versions of (a) the Cambridge dictionaries, where they will also tell you which are the British and which the American spellings,[7] or (b) Webster's online with input from Princeton and INSEAD (*Institut Européen d'Administration des Affaires*) professors.[8]

4. Guidebooks from other scientific disciplines and laboratories that follow the same principles as A&A so you can use them for most questions, even if their examples come from their disciplines. These include

   - CHEMISTRY: The ACS Style Guide: A Manual for Authors and Editors, Second Edition,[9] edited by Janet S. Dodd. Its sections are: Getting Started, Writing Style and Word Usage, Components of a Paper, Types of Presentations, Advice from the Authorities.

   - BIOLOGY, the biology department at Columbia University "WRITING A SCIENTIFIC RESEARCH ARTICLE."[10] Its main sections are found under "Format for the paper" and "Edit your paper!!" Both of these repeat in detail what language editors are looking for when editing your papers, so you can see there that they are not being any more rigorous than others are.

   - NASA – Langley Research Laboratories guide for authors presenting reports by Mary McCaskill.[11]

---

[7]http://dictionary.cambridge.org/
[8]http://www.websters-online-dictionary.org/
[9]http://www.oup.com/us/samplechapters/0841234620/?view=usa
[10]http://www.columbia.edu/cu/biology/ug/research/paper.html
[11]http://www.sti.nasa.gov/publish/sp7084.pdf

5. The latest edition of *The Chicago Manual of Style*, Chicago University Press. This extensive manual is used by most scientific communities in the United States, including psychology. It may be too technical and detailed, when most of what you need is on the A&A site. The book itself is bulky, but it is possible to subscribe to it online for $ 30 (US) a year to access all its sections (including the style guide) and even to ask specific questions of its staff of editors.

6. Other resources online for scientific writing and writing in general:

   - Madison Wisconsin Writing Lab for Scientific Reports,[12] including suggestions for each of the 6 parts of a paper (or scientific report).

   - A list of common errors and advice by a professor of English at Washington State University, Paul Brians. It tells you it is meant for native English speakers,[13], but there is much that can help everyone. He also gives a list of sites specifically for second-language English writers.

   - Purdue University Online Writing Laboratory (OWL) pages for second language students of English.[14]

   - English-at-home.com (go to "grammar").

6   Concluding statement

The search for coherence and quality includes approaching the style that is considered to be best in the language being used, and in English this is best expressed by Strunk & White's *Elements of Style*. If science writing considered that using the jargoned clichés from its informal style were better than plain and standard English, then the style guides for the various scientific disciplines would not refer back to this reference work over and over, as done here.

At times while working through a manuscript I empathize with authors who had first learned a correct and clear English to express their findings. As time went on, however, they picked up these phrases in the literature and at conferences, learned what they meant, and supposed that these were more appropriate for some reason. Since scientists use them, they must be more precise; or perhaps since an English speaker uses them, they must be more correct, a writer might think. Many are correct, admittedly, since some or most are found in a dictionary, which is appropriate, but most of them are either bad style or hard to use correctly.

The work that has already been put into the papers is considerable before it gets to the language editors, and the English is usually clear enough to make suggestions to make it more effective, as well as more correct. If this work on the English had not been done by authors before submission, little could be suggested beyond basic grammar and vocabulary corrections alone. Going beyond this first level to suggest changes for clarity,

---

[12]http://www.wisc.edu/writing/Handbook/ScienceReport.html
[13]http://www.wsu.edu/~brians/errors/
[14]http://owl.english.purdue.edu/handouts/esl/eslstudent.html#purdue

consistency, and a smooth style is intended to move in the direction of improving the impact of the research published in A&A.

The response to language editing from the majority of the authors who responded to the Author Survey of 2007 supported the Journal's efforts, just as most authors respond very positively to these efforts by either incorporating the suggestions, by explaining and communicating constructively when not understanding or agreeing with some of the suggestions, or even by sending information on technical expressions that helps the language editors in their task. Communication is, after all, the goal from the time the observations and calculations are completed to when this work and its results need to be expressed and disseminated to colleagues at home and abroad.

This work is based on all the collaborative efforts of the language editors and staff at A&A through the years. Special thanks go to Claude Bertout and Claire Halliday for their useful suggestions on the final draft.

Scientific Writing for Young Astronomers
C. Sterken (ed)
*EAS Publications Series, **49** (2011) 85–105*
www.eas-journal.org

GUIDE TO THE ENGLISH EDITING
AT ASTRONOMY & ASTROPHYSICS

J. Adams, C. Halliday, A. Peter and M. Usdin[1]

Abstract. We reproduce here the guide to the English editing currently in use at A&A.

1   Aims of this guide

This language handbook can be used in different ways, which include finding explanations behind some of the suggested changes. It was written based on the kinds of changes we recommend most often in A&A papers, so it does not pretend to be a full English language guide.

The most useful time to use this A&A handbook is while working through the changes we have suggested in your accepted article, whenever you are not able to understand why a change was suggested. It not only contains details, but also tries to give a sense of the spirit behind certain types of changes we ask for, from some simple conventions about details of spelling or punctuation, through causes of ambiguity, to changes for the sake of rhetorical effectiveness. At times, for instance, some of these explanations will have been inserted into the yellow note boxes in your corrected version, when a problem occurs often in a paper.

A second use might also be to take a look at it before submitting a paper so as to anticipate some of the changes we will suggest anyway or, alternatively, as a "style guide" to supplement any you might be using at present.

2   Consistency of style: UK/US spelling conventions

A basic principle of our editing is to ask for consistency within an article, whether in the punctuation, capitalization, spelling, or abbreviations. One example is found in words that have more than one proper spelling or form: **disc/disk, online/on-line/on line**. A good dictionary will tell you both versions of any word, and some tell you which is more common to the dialect you have chosen.

---

[1] Observatoire de Paris, 61 Av. de l'Observatoire, 75014 Paris, France

This dictionary should also let you know which spellings are American (US) and which are British (UK), when there are differences. There is only a small difference between the two conventions, but you need to use the same one throughout a paper. Use the following guidelines when asked to revise the spelling for this reason, and of course your text processor (including LATEX) can track this for you.

UK SPELLING CONVENTIONS	US SPELLING CONVENTIONS
-OUR endings: Behaviour, neighbour, favour, colour, harbour, vapour	**-OR endings** (pronounced /or/): behavior, neighbor, favor, color, harbor, vapor (NB, but "contour", pronounced /oor/)
-RE ending: centre, metre, fibre, calibre	**-ER ending:** center, meter, fiber, caliber
-SE ending on some verbs (less often for scientific terms) or **ISATION** analyse, summarise, organisation to practise (noun = practice)	**Usually -ZE ending for the same verbs** or **IZATION** analyze, summarize, organization to practice
-DS/ST ending: towards, amidst, whilst, backwards	**D ending:** toward, amid while, backward
Doubled consonants on verb endings tend to be British, but are not strictly so: Modelled, labelled, travelling	**US English tends to the single consonant** for endings if the pronunciation does not change. modeled, labeled, traveling
Hyphenation: on-line, non-zero north-east, south-west	**Merging** online, nonzero northeast, southwest
Other words & expressions: of the order of, in the order of Brackets & square brackets Further (or farther if distance) Autumn, artefact Nought, grey Ageing, speciality Catalogue, analogue, haloes, to orientate Fulfil, sulphur	**Others:** on the order of (use the standard synonyms: roughly, approximately, about) parentheses & brackets *Stricter on the difference* fall, artifact zero, gray aging, specialty catalog, analog, halos, to orient fulfill, sulfur

3 Punctuation, abbreviation, and capitalization

As for all rules, some of the following are conventions chosen for stylistic consistency within A&A. Most, however, are aiming to make the text clearer for readers, so anything that causes them to stop in order to understand or puzzle through a sentence may also make them stop reading altogether and go on to something else, which is neither the goal of the Journal nor the aim of any author.

How to write dates

In the text of your article, you may use either date formats common to regular prose (Day Month Year or Month Day Year) or the IAU recommended format (Year Month Day), as long as you remain consistent. Please use letters for the month, rather than using its figure, to avoid ambiguity.

> When the date is an integral part of the event's name, the use of the IAU format is recommended but not mandatory (for example, "the 2003 January 17 CME event"). Dates included in tables should be in the IAU abridged format (*e.g.*, 2003 Jul 4).
>
> Otherwise, you are free to choose from the following formats, as long as the format is consistent: 3 December 2005 / 3 Dec. 2005 / Dec. 3, 2005 / December 3, 2005 / 2005 December 3 / 2005 Dec 3.

3.1 Punctuation

The serial COMMA in lists

In most technical and scientific writing, a comma precedes the "and" or "or" before the final item in a list of three or more[1]. In some cases, we may suggest a comma where you consider it inappropriate, but that probably shows that the list is ambiguously formulated, so it needs to clearly indicate which items are in the main list and which words modify those items.

The simple form is **X, Y, and Z** or **X, Y, or Z**. When any one of these is a compound, then the comma becomes essential: **W, X and Y and Z**. Should it be **W, X and Y, and Z**, where X and Y work as one element in the list? Or is it either an incorrectly punctuated list of four items (**W, X, Y, and Z**) or else only the last two work as a single unit: **W, X, and Y and Z**?

LISTS after colons

A colon is an appropriate way to introduce a long or a complicated list, and most authors do this correctly. No colon should follow the first colon in a sentence, and the sentence must end at the end of that list.

To separate the items in the list, **use commas when the items are simple and semicolons for those that are complex**, which means that one or more of the elements require a comma because of a list or dependent clauses. Also, you need to keep each item parallel to the others (same part of speech or similar syntax, which is called "Parallelism" in language guides, see Sect. 5.5 below.).

---

[1] The purpose is to avoid confusion, although many debate this convention (see Wikipedia at http://en.wikipedia.org/Wiki/Serial_comma).

> A simple example is
> "…as follows: the Earth, the Solar System, and the Galaxy."
> The complex formulations look more like
> "…as follows: the Earth, but not the Moon; the Solar System and all
> its planets; and finally the Galaxy, which is a member of the Local Group."

Like the serial comma above, the combination of comma and semi-colon makes it clear which elements go in which part of the list. One place where colons are not needed is when introducing an equation that is separated in the layout but continues the syntax of the sentence, *e.g.*, after a verb or a preposition, where no punctuation is needed. In fact, it would be incorrect in a normal sentence layout, so too here.

COLONS (or semi-colons) between two full sentences

That any two full sentences are connected in the original shows that the author finds a relationship. When there is a relationship, then a word is more precise than the ambiguous punctuation. Most often, this word is a causal connector, such as "because" or "since". It is possible to use colons and semi-colons to connect two sentences once in a while when the relationship is obvious to all, but make sure that there are not too many in each paragraph, which indicates a need to replace them with the logical connectors to help the reader follow your thought.

Obviously connecting two sentences by only a comma (no logical connector) poses the same problem and is technically incorrect in English; but to be honest, there are very few authors who submit papers to A&A with this style error[2].

> The preferred uses of the semi-colon between two sentences are either
> (i) when one of the sentences, at least, is complex and contains one or
> more commas or
> (ii) where a connector that requires a comma after it (however, moreover) is
> used. In contrast, the colon tends to be used mostly to introduce an example
> or a list.

COMMAS between two full sentences (each with its own subject and verb) with a connector

The lack of a comma between two full sentences can be confusing at times, and it is needed for longer compound sentences, unless a semi-colon is more appropriate. Please consider that this is usually recommended in technical writing guides, except in very short sentences where there is no ambiguity as to their relationship.

---

[2]See "comma splice" or "run-on sentence" in English guidebooks.

COMMAS after introductory clauses

A comma is usually needed after a subordinate introductory phrase and the main clause, especially because it is often difficult to tell where the main clause begins, even for specialists. When this occurs with figures, it requires a lot of attention by the reader ("In the interval 10 K < Tex < 30 K, F(Tex) is ..."), but is often nearly impossible when following a long string of nouns that are used as modifiers, in such a way that the final true noun and the first noun of the main clause become nearly undetectable.

> Example: "As previously emerging radiation spectra are calculated for a radius of the scattering cloud equal to the Bondi radius." Does the author mean: "As previously, emerging spectra are calculated"
> or: "As for previously emerging radiation, spectra are calculated";
> or even, is this a fragment that introduces the next sentence?

If a specialist in this precise field might have to hesitate to decide between the two (or more) choices, one not a specialist will have to look much harder, will not understand, or will give up. A comma solves the problem for all cases and keeps your reader interested.

More often, authors invert the normal order of the English sentence, which is acceptable if not used constantly. There should be a reason for it, say, for transition or rhetorical emphasis, rather than simply for varying the syntax.

COMMAS for nonessential sentence parts

The nondefining relative clause (dependent sentence part) must be separated from the main sentence by commas, as in "Galileo, who had a daughter named Virginia, discovered the phases of Venus." Do not precede the relative pronoun "that" by a comma, because it is only used to introduce a defining clause, unlike "who", "which", etc., which are used for both. (*cf.* Sect. 5.6.1).

In contrast, the defining relative clause is an essential part of the main idea in a sentence, so it is never surrounded by commas, as in "This method works well if it is applied consistently." This implies that the method will not work well if not applied consistently, and consistent application is also the main idea here.

> Examples: "The Galaxy has a thin disk that is rich in metals." (This means that metallicity is an essential property of the Galactic thin disk and is the topic of the discussion at hand.)
> Compare with "The galaxy has a thin disk, which is rich in metals." (The nondefining clause adds an interesting detail, which might not be essential to the Galactic thin disk, but is definitely not essential to the discussion where the sentence occurs.)

HYPHENS for two-word compounds when used together as an adjective

The rule is that the hyphen is used to connect two words, *e.g.*, an adjective and its noun or two nouns, when together they are used to modify another noun: "a region where stars are forming" becomes "a star-forming region". However, it is not a hard-and-fast rule, but instead depends on whether there is a precedent for that spelling or else whether the hyphen helps to clarify relationships that are otherwise ambiguous. Also, notice that the modifying noun becomes singular[3]. The compound cardinal directions are not hyphenated in US English: **northeast, southwest**.

Examples of ambiguity:

- "long time frame observation" is ambiguous without a hyphen, because both the timeframe and an observation can be long. A "long-time frame observation" is an observation of frames over a long period of time (better without the second part of the hyphenated phrase, obviously), or "long time-frame observation" where it is essential that the observation is made within a specific, defined frame. Consider also "high mass star forming cloud". Do we mean either (1) high-mass star-forming cloud (a star-forming cloud of high mass) or (2) high-mass-star forming cloud (a cloud forming high-mass stars? It is better that the author controls how the reader understands the relationships between each part of the phrase.

- Another example of this might be "The local limb slope", where a non-expert might wonder if the limb is local ("local-limb slope") or if it is a limb of a slope ("local limb slope"). According to the rule, it must only be the second choice, since there is no hyphen.

The hyphen is particularly useful in complicated compounded phrases that also have these compounded noun phrases: *e.g.*, "Closer inspection of our direct mass and luminosity function determinations reveals..." Are we to inspect the direct mass more closely and also the luminosity-function determination? In these cases, as in so many, we may suggest the hyphen, which is a clue that there might be a reason for it in that situation. You are asked to consider its usefulness in that context; so when you feel it is clear for all astrophysicists, even those starting out in the specialty, then it need not be inserted in this case.

PARENTHESES/BRACKETS

We can suggest that long sections in parenthesis be placed in the running text, where they are more appropriate than in, say, a footnote, because it relates to the subject at hand.

---

[3]See "Comment" on this http://grammar.ccc.commnet.edu/grammar/compounds.htm

> Examples:
> "We observed these stars for 24 hours (which did not include breaks to eat and sleep)." ⇒ "We observed these stars for 24 hours, which did not even include breaks for eating and sleeping."
>
> You may use the same sentence but as a second parenthesis, punctuated as "We observed these stars for 24 hours. (This did not even include a break for eating and sleeping.)"

Avoid parentheses within parentheses, except when it is part of a formula. This includes the year in a reference that is placed in parenthesis, where we also prefer no comma between the name and year, but do expect the period after " *et al.*": (Johnson *et al.* 1999). When the name is in the text, then only the year is placed in parenthesis.

3.2 Capitalization and abbreviation

- A&A asks you to **capitalize only the first word of the title and subtitles**, with the exception of proper nouns. Abbreviations are to be avoided in titles, unless they are very common ones such as cardinal directions (NE), some star names, or chemical terms.

- Like other journals, A&A asks you to remain in the formal register; that is, do not contract as in "don't" or "can't", while the second is one word: "cannot".

- Follow the Instructions to authors at `www.aanda.org` about capitalizing and abbreviating numbered references in your sentences, such as Figure 1, Section 2, Eq. 3, or even Col. 4. When possible, use the plural for multiple items: "Figs. 2–4" or "Eqs. 5 and 6". "Table" is never abbreviated, although it is capitalized when followed by its designated number, while no word is abbreviated at the beginning of the sentence. Also, when any of these words are not followed by a number or letter indication, then it is treated as a normal noun that is not capitalized.

> Example: "This argument can be found in the next section, where Eqs. 5 and 6 are explained."

- **CAPITALIZE all adjectives and verbs formed from proper names**, such as Poissonian or Compton, no matter how familiar it has become: Newtonian. One of few exceptions to this is "cartesian".

- A&A prefers lower case for all generic names and terms (*e.g.*, spectral energy distribution, active galactic nucleus), but the upper case is normal for proper names and generic ones that are traditionally used for a single, special object (*e.g.*, Local Group, Magellanic Clouds). A combination of both rules is found in the generic "hot Jupiter" and others like it. Finally, the upper case is used for copyrighted material, unique instrument names, and observatories.

- Unlike the plural and lower-case S in abbreviations of names (acronyms) and numbered references (Sects. 5 and 6), abbreviations of measurements (yr, mg, m, s, etc.) do not add an S.

- **Use the LOWER CASE when introducing abbreviations and acronyms.** The words that make up the introduced acronym are not capitalized unless either a word is a proper name or the acronym does not use the first letters.

> Examples: star formation (SF), Atlantic Ocean (AO), the Galactic center (GC), HIgh-Precision PARallax COllecting Satellite (HIPPARCOS)

- Write out numbers below eleven in full unless they are connected to specific measurement ("8 Gyr") or are used with another number above ten ("between 8 and 20").

- Use the lower case for the full names of chemical elements and for the cardinal directions: lithium, southeast.

4   Verb tenses, plural/singular nouns

4.1   Verb tenses

The present tense

The present tense is used for statements of fact and general truths, for a set of steps in a method (not for the steps in testing), and for any results, discussion, and conclusions that are set out in the paper itself. We do accept the present perfect and present simple for the work of others when the paper is cited, which comes from etiquette learnt for this situation. It must be consistent, however, and should avoid any time tags that contradict that tense choice.

To be consistent, avoid using the present for your work and then the past tense for others' work in their papers, because it may suggest theirs is wrong.

The present perfect

Using the present perfect in English implies that the situation, even the action, is still going on, so not completed. It can also be used to put strong emphasis on how the situation has changed just this instant, usually with the negative: "I've never read such a fine article as this one." This occurs because the present perfect is used in other languages for a finished, completed action, so we usually take this use as intending the simple past in English. However, we can obviously be mistaken in some cases. It is, however, true that most English style moves quickly out of the present perfect and avoids using it throughout a full paragraph in sentence after sentence, unlike in other languages. It is perceived as wordy.

The future

The future is expressed in English with either the present simple or the verb "will" and the infinitive. **When referring to work that follows in the same paper, use the present simple as appropriate to the immediate and certain future**, while the classical form with "will" is used for future work after this article or for predicted events. It is used for events that are relatively certain, otherwise one of the conditional forms might be needed. "We explain our method in Sect. 2." *vs.* "Once these stars have been observed, they will be analyzed using the new method."

Active *vs.* passive voice

Scientific language tends to use the passive, especially when talking of work done in a team; however, this has changed with time, so that **A&A encourages authors to use the more personal and active forms when possible**. Best for style is to vary the two and to use the active when the passive requires many more words.

> Example: "The fact that these observations can lead to accurate results is demonstrated by our analysis."
> ⇒ "Our analysis demonstrates that these observations lead to accurate results."

4.2   Nouns: plural or singular

Some nouns can be either singular or plural depending on their context, although each has a preferred number.

NUMBER
    When you are referring to many individual things, use the plural form to mean "many stars" not "the total number of stars"; *e.g.*, "A number of stars **were** found in our last observing run." Compare this with "The number of stars exceeds several thousand."

MAJORITY
    "Majority" and "variety" follow the same rule as "number", although the sense of the word is the determining factor: do you mean it as a single entity or as a group of individuals? If you use this phrase, then make certain it is truly over 50 percent and just "many".
    PLURAL: "The majority of scientists hope the agency will raise the number and amount of grants."
    SINGULAR: "The overwhelming majority still votes for amnesty."

STATISTICS and others like it
    The words statistics, systematics, and dynamics are usually plural when they refer to a set of measurements: "The statistics for our sample of stars are found in Table 1." If you want to make it singular in this meaning, then use "set of statistics". Each may have a singular use, too, to mean a single one of these measurements.

When referring to the study of statistics, dynamics, or systematics, then it is also singular, but this is a rare use in astrophysics.

DATA

Like other scientific publications, A&A remains "purist" by asking for the plural usage[4]. If you mean a single one, use "data point" instead of "datum".

5 Structure: Sentences and word order

5.1 Paragraphing

We do not usually have time to deal with the most effective way to develop an idea in full paragraphs, because we are more concerned with more basic English problems related to clarity, but it is a concern and will be addressed when it is clearly a problem in any article. There are times we make suggestions for paragraph breaks or ask an author to combine a single sentence swimming alone with one of the neighboring paragraphs.

Otherwise, the best choice for paragraphs, as for sentences, is to aim for variety in length, if possible. Having too many of the terse two to three sentence paragraphs in a row leads to monotony and a lack of transition or logical movement within a section, which is just as true for a series of short simple sentences in a paragraph. Paragraphs that are either too long or too short can be a problem for different reasons, with different solutions[5]. Each paragraph should have its own focus that is introduced in the first sentence and that includes some brief transition phrase from the preceding paragraph or else a clear connection to the topic of the whole section. This topic sentence should then be followed by two or more other sentences that develop and prove that idea, *i.e.*, the standard expository paragraph.

5.2 The English default structure

The English declarative sentence has its main ("default") structure of subject (noun phrase), predicate (verb phrase), and objects of the action. It can be varied, if there's a reason, but should be used as often as possible. Here, though, we mention those kinds of errors that we see often and that lead to either ambiguity or else to awkwardness. Avoid separating the subject from its predicate (the main action/verb of the sentence): *e.g.* "The author, after revising the English, submitted his article." becomes "After revising the English, the author submitted . . .".

---

[4]For more examples of where the sense of the noun and intent determine number of the verb after an apparently singular noun, see Fowler at http://www.bartleby.com/116/202.html

[5]See the short Wikipedia comment on both at
http://en.wikipedia.org/wiki/Wikipedia:Guide_to_writing_better_articles#Paragraphs.
And see advice on the expository paragraph on "Commnet" or in any style guide you may have.

5.3   Inverting word order

Word order is often awkward in papers we see, because a phrase has been placed in an unnatural position or just because a word is out of its normal place in a sentence. There must be a good reason for inverting the order, and such a phrase should be separated by a comma in most cases.

EXAMPLE: "Especially appealing is the study of its stellar winds." ⇒ "Study of its winds is especially appealing". Unless this comes at the end of a rhetorical list we seldom see in scientific writing, then there is no reason for this inversion and the only improvement needed to the second version is to perhaps get rid of the "of" phrasing: "Studying its winds is ..."

5.4   The direct object

The direct object goes immediately after its verb with rare exceptions, and should not be separated without a very clear reason: "The star's image shows clearly the alignment." ⇒ Use either "clearly shows" or "shows the alignment clearly." Avoid interjecting long subordinate phrases as adverbials between the transitive verb and its direct object or in the middle of any clause, for that matter.

And of course a transitive verb should always have a direct object, with the most familiar problems "to allow", "to enable", and "to permit".

> The error in "This program allows to analyze the data" can be solved in three ways:
> (i) "This program allows us to analyze the data",
> (ii) "This program allows the data to be analyzed", or
> (iii) "This program allows analysis of the data".

Several adverbs go before the verb: "only" and "also" are two familiar examples. But **there are many preferred places for different grammatical forms, as in all languages, so if one has been corrected, please check a more basic grammar guide.** If you felt that there was a reason for moving from the normal order, then perhaps something else in the sentence kept this emphasis from being clear as we were reading.

5.5   Parallel structure

Parallel structure is essential for avoiding ambiguity in compounded phrasing, whether in a list or between two longer sections of a sentence (See "Parallelism" in style guides).

The compounded (or listed) items should share the same part of speech or syntax and all have the same grammatical status.

> **EXAMPLES**
>
> "Emission in the lowest CO rotational transitions was detected at the position of V838 Mon and nearby". ⇒"Emission in the lowest CO rotational transitions was detected both at the position of V838 Mon and in its close vicinity"
>
> "UCAC3 provided us with a large data set that has suitable coverage in proper motion and with reasonable errors for our study" (direct object and prepositional phrase) ⇒ "UCAC3 provided us with a large data set that offers suitable coverage in proper motion and sufficient accuracy for our study."
>
> "It is best to use alternative methods because of the obscured target and because this is always a good idea in astronomy." (mixing a noun with a full clause) ⇒ "It is best to use alternative methods because the target is obscured and because this is always a good idea in astronomy."

Many elements are often combined in lists in the very long sections with details of scientific work and results. **The rule is to make certain that the related items in the list maintain the same part of speech and syntax, which includes repeating the preposition or obviously reusing the structure of the first in the others.** When faced with several lists in one sentence (even compounding within lists), the logic becomes clear, for otherwise the reader cannot sort out which goes with which.

5.6 Unambiguous reference

Although it happens to the best of us, it is rare that we need to remind writers for A&A to check that an antecedent is clear in normal sentences. It does occur, however, in more complicated constructions, so the following rules were set up to avoid creating ambiguity and awkwardness in reference.

5.6.1 Subordinate clause: THAT *vs.* WHICH

In scientific journals, we prefer that the relative pronoun form of *THAT* is used for all defining (restrictive) connections and always without a comma, while *WHICH* is used only when the information following is added and not essential (nondefining) to the main idea of a sentence, something many are tempted to put into parenthesis, for instance, and is therefore always introduced with a comma after the preceding word[6]. See also page 89.

5.6.2 Subject of a subordinate clause: Dangling modifiers

And of course **make certain that the subject of a subordinate phrase is really the same as the subject of the sentence** to avoid the "dangling modifier" that can be amusing at

---

[6]The King's English at http://www.bartleby.com/116/204.html.

times: "Having studied the spectra, the two galaxies were approaching each other" ⇒ "Our study of the spectra showed that the two galaxies were approaching each other."

Any other form is ambiguous. Avoid gerunds unless they are used as a noun phrase alone ("Studying stars is my job") or in a prepositional phrase ("After reading the ..., I ...").

Examples:
"An ordered component to Lz is possible, **producing** a net planetary spin angular momentum." Is it "because it produces" (cause), or "thereby producing" (result)?

"Here we analyze a quiescent prominence with fine structures **exhibiting** dynamic behavior, which was observed ..." Is it "a prominence with fine structures, which exhibits ..." or "a prominence with fine structures that exhibit ..."?

---

6   Clarity and precision

In English, it is a stylistic virtue to be concise and to use the precise phrasing for an idea[7], while in science anything that is precise and clear is a necessary virtue, so that on this the demands of language and science coincide. In scientific writing, however, many wordy constructions have developed for various reasons and can be tolerated until too many of them combine to cloud the meaning. Likewise, poetic language and puns that work so well in other contexts should be avoided in scientific articles as much as possible, especially when so many non-native speakers cannot appreciate them. On the other hand, some metaphors have entered into the terms of some specialties and other expressions are so common that they have become clear to close colleagues, but not necessarily to the whole community.

Active phrasing

One way to do this is to aim for ACTIVE PHRASING as often as possible for several reasons, such as clarity, lively style, avoidance of too many "of" preposition phrases and "-tion" nouns. We look for any overuse of abstract nouns where the original verb form is more appropriate: "the estimation of x is ..." should be "estimating x is..." "We observe the planet by the detection of the stellar light it reflects" ⇒ "We observe the planet by detecting the stellar light it reflects" or even better ⇒ "We detect the planet by the stellar light it reflects." Considering that science requires these nouns, using more than is required means that it quickly becomes too abstract and wordy.

---

[7]References include "Commnet" at http://grammar.ccc.commnet.edu/grammar/concise.htm.

6.1 Indirect introductory phrasing

Indirect introductory phrasing should be used sparingly by the same token, as in the following examples.

- Please avoid "It is worth stating that" or "We want to stress that" most of the time, if not always. In most cases it takes away from the strength of the assertion that follows, and is understood at any rate.

- Avoid "we hope", "we wish", "we believe", etc. Even "we can see that ..." or "we think that ..." are already assumed by the reader, so are not useful.

- Do not start a sentence with "concerning" or "with regard to": *e.g.*, "Concerning the value of the frequency, we have investigated whether stars release ..." should be "We have investigated the value of the frequency at which stars release ..."

6.2 Avoiding ambiguity

Avoid ambiguous wording whenever possible, as can be seen in "we increase the statistics." Does it mean either "raise the quality" or "increase the number of statistical tests"?. One notorious example is the connector "and", which in English is used too often for the more precise connector that is meant: "Divide and conquer" for "Divide, then conquer". **The most obvious source of ambiguity comes from words that have more than one meaning or use**, especially for connectors that are also used as adverbs, prepositions, and adjectives:

- **like**   do you mean "such as"? or "similar to"?
- **as**   do you mean "because" or "while" or "during"?
- **since**   do you mean "because" or "after which"?
- **while**   do you mean "whereas" or "during"?
- **so**   do you mean "therefore" or "meanwhile" or?
- **quite**   do you mean "very"? or "somewhat"?
- **rather**   do you mean "instead"? "very"? "somewhat"?
- **further**   do you mean "more"? "another"? "again"? "an extended"?
- **such**   do you mean "this sort of"? "very much"? "these"? etc.
- **few**   do you mean "very few" or "a few" or?
- **due to**   do you mean the adjective (default use)
          or an adverbial meaning "because of", "owing to" or "thanks to"?

Example: "The flat-fielding correction is a delicate step at near-infrared wavelengths due to the fringing effects." This says that the wavelengths are caused by the fringing, so if this is not the meaning, then it needs rephrasing to "The fringing effects make correcting the flat-fielding delicate" or something with "owing to".

Or from words that are close in sound, and often in meaning:

- insure/assure/ensure
- than/then
- affect/effect/impact
- relative to/compared to/with respect to
- comprise/consist of/be composed of
- in contrast to/contrary to/opposed to/compared to.

Gerunds (V + ing) can also lead to ambiguity, because they are used in so many different ways: present participle (A debate is going on), adjective (ongoing debate), connector to indicate both cause (Having won the Nobel Prize, I celebrated) and effect (I won the Nobel Prize, thus having a reason to celebrate). See the example used above in Section 5.6.2

Another major source of ambiguity is reference confusion (see Sect. 5.6), but also simply the use of too many words where a simpler phrasing is more appropriate because clearer. Likewise, non-native writers in English might not be able to sense when a longer phrasing is appropriate or not, not that all native English writers do it well all the time either. Some areas where problems arise regularly in our experience follow.

The first and easiest is to avoid any sexist or hemispheric phrasings such as "Summer 2005" (give the months rather), and if you do not know the referee's name then please acknowledge "her/his" or "the" useful comments, not merely "his". This occurs less in astrophysical writing, but do try to avoid using "American" alone when you mean "North American" or from the "United States of America".

In general, **try to use the fewest number of words possible, which often makes your ideas clearer,** as long as the reader has a better chance of understanding your text. However, too much compactness can also cause confusion, because the connecting words, especially prepositions, have been removed. This happens a lot in scientific compositions of nouns modifying other nouns to avoid using the prepositional phrase with "of": "a theory of how stars are formed" becomes "a star formation theory".

> As an example of compacting too far, we see phrases like "high statistics"
> for "high-quality statistics". Writers often leave off the "that is"
> in modifying phrases and do so quite correctly, but it can also go too far,
> so we may suggest putting it back in to avoid ambiguity or for clearer style:
> "a theory more interesting than the facts" ⇒
> "a theory that is more interesting than the facts".

6.3 Avoiding redundancies

These are examples of using two words that both mean, or at least imply, the same thing.
They are wordy, though many have entered colloquial language (*e.g.*, "slim and trim").
Ones that we have found during our editing include:

- coupled together ⇒ coupled

- the obtained results ⇒ results

- determine the point of onset of... ⇒ determine the onset of

- decreases down ⇒ decreases

- such as *e.g.* ⇒ such as

- such as ..., etc. ⇒ such as (or for example)

- effect due to ⇒ effect of

- on the order of a few ⇒ a few.

6.4 Avoiding vague qualitative descriptions

- A good example here is "rather small". In this case, compare it with something larger.

- In English, if not most languages, "very important" means the same thing as "important", and even less when "very" is overused.

- Likewise, watch out for undermining absolutes: if only "partly true", then which part is true and which not (?); "somewhat consistent" ("consistent" is absolute, so it cannot be partly so).

- Or else "in fact" or "the fact that" need to be avoided as often as possible; make certain it is a fact.

- "In any case" and some other familiar filler expressions cloud scientific writing, so use them when you actually mean it. This example does not mean "Oh well!" in writing, unlike in discourse, or the more recent Americanism "Whatever!" If there are two possible cases being discussed, use "in either case".

- Misuse of "well" is frequent, although some technical phrases are normal (well-constrained). "It is well visible" is both incorrect and vague, so use "It is visible" or "clearly visible". "It is well understood" is vague and awkward to mean "fully", "nearly", or perhaps "completely".

6.5 Avoiding wordy constructions that mean the same as the direct one

"In order to" can usually be shortened to the infinitive alone "to" plus the verb.

"On/of/in the order of" and "within an order of magnitude of" are equivalent to the more standard "approximately, about, roughly", and are by no means more precise. Try to use the other forms as often as possible and avoid using redundant forms here (on the order of a few). They are never used as a substitute for "the same as" or "similar to", unlike in other languages.

- X is of particular importance ⇒ X is particularly important

- X is only of an approximate nature ⇒ X is only approximate

- To detect the presence of an atmosphere ⇒ to detect an atmosphere

- To perform an identification ⇒ to identify

- Analyses were made ⇒ We analyzed

- To make adjustments to ⇒ to adjust

- To give consideration to ⇒ to consider

- It is shown in Figure 2 that ⇒ Figure 2 shows that

- Concerning the explosion, its effect is ... ⇒ The effect of the explosion is ...

- The intersection of x and y occurs ⇒ x intersects y

- We aim at estimating (If not in the abstract) ⇒ We estimate

- Make a comparison with ⇒ to compare .... (Likewise other noun versions of active verbs, when not needed for meaning)

- Due to the fact that ⇒ because

- Show strong indications of something ⇒ to indicate something strongly

- Is in contradiction with ⇒ contradicts

- Is found to be in agreement with ⇒ agrees with

- Before proceeding further, it is worth commenting at this point that we have studied the ... ⇒ We have studied the ...

And finally, stay away from the double negative in favor of a direct, affirmative statement: "This result is not unlikely" can be either "This result is likely" or "This result is possible". Notice that the "litotes" format is slightly ambiguous, because it does not indicate the degree precisely, except for what it is not.

6.6 Transitional phrases

Indicating transistion is useful, but when they only add words or repeat themselves, they need to be avoided: **Avoid "actually", "in fact", "moreover", "indeed", etc., or use only when they help clarify relations** rather than muddle them. Likewise, "hence" can be very useful in the right context, but when overused becomes useless and loses its rhetorical power. Once or twice a paper makes sense, but not once or twice a page or more often. Even if tempted to place one of its synonyms, be aware that overuse of several of these in a row begins to seem odd, too, just as using a string of sentences all containing indicators of contrast (but, however, in contrast, etc.) seems like the writer is arguing with her/himself.

7 Some frequent corrections

This section discusses several words or phrases that need to be dealt with regularly because they are either incorrect or vague.

• Overuse of "SUCH" for "THIS" or other choices

Example: "such stars" after already referring to specific ones ⇒ "these stars"

   "SUCH" is not the normal demonstrative article, and its normal meaning ("this kind of") is not what is seen in most non- English-speakers' papers. When this meaning of "such" is used more than one third as often as the equivalent "this", then it needs changing except for the case of "this sort of" or "this kind of". Also, watch out!-"such a" can be insulting in some contexts when the author maybe meant only "this": *e.g.*, "... (Jones 1994). Such an article is ..." It is also ambiguous is several instances, and the author will be asked to choose either "this" or "this kind of" to eliminate the confusion for both English native and non-native readers.

• Overuse of "FURTHER"

"Further" is often used exclusively for the standard "more" or "another/other" or "earlier/former" or even "then" in a sequence (!), *e.g.*, "further examples" for "more examples". Each of these means something slightly different from the others, and sometimes means something truly different, leading to ambiguity.

   Examples:

   • adds a further layer to ⇒ adds another layer to ...

   • We did further research ⇒ We did more research (or We continued our research)

It means "more extended", not simply "a greater number of" as so often seen.

- Use of "THESE/THIS LATTER"

"This latter" is redundant, and "latter" is generally overused in the articles. "these/this latter X" ⇒ Either "these/this X" or "the latter X", though the first is preferred. "The latter" only for two, "last" is for more than two, even when you mean the last two of only three. You are welcome to use the standard "first" and "second", which mean the same and are just as precise.

- Overuse of "TOGETHER WITH" and "AS WELL AS"

These phrases can be very useful structures in English when used properly, but they are often overused in articles we see. "Together with" often replaces "with", and "as well as" is used too often for "and" in a list. Only "and" is used after "both", never "as well as", as in "Both X as well as Y are ..." ⇒ "Both X and Y are ..." In a compounded subject with either of these phrases, the verb agrees with the main noun(s), not with what follows these subordinators, which are enclosed by commas.

> Example: The Sun, together with all stars with planets, have ... ⇒
> The Sun, together with all stars with planets, has

- Overuse of "BEST"

As for "well", use the precise word for what is meant. We suggest a choice of synonyms for "best", all except "favorite": precise, useful, clear, most likely, etc.

- High/low *vs.* large/small

"High" and "low" are used more often in English than in other languages that seem to prefer "large" or "small" for the same idea. English also uses "greater" much more than "larger and "less" or "fewer" more than "smaller", just as it uses all the standard comparative structures (more than, less than, fewer than). N. B., some examples below show up in two places.

- **High or low** for most words that are typically indicated by up/down graphs: abundance, absorption, accuracy, adiabaticity, background, brightness, contrast, degree, density, energy, extinction, fraction, frequency, latitude, level, luminosity, mass, metallicity, number, percentage, precision, pressure, probability, proportion, quality, rate, redshift, resolution, shear, speed (fast, slow), temperature, value, velocity

- **Steep or shallow:** gradient, slope

- **Big/large or small:** amplitude, broadening, diffusivity, momentum, opacity, proper motions, radius, uncertainties

- **Can be either high/low or large/small:** It depends on the emphasis: dispersion, frequency, magnitude, number, opacity

- **Short or long:** burst, length, period, time, timescale

- **Strong or weak:** acceleration, anisotropy, argument, assymetry, constraint, contrast, current, dependence, evidence, magnetic field, flow, gradient, instability, pulse, relevance, shear, turbulence, velocity field

- **Wide, broad, or narrow:** range, spread, variety

- **Tight or loose:** Often the same as "strong/weak": correlation, relationship

- Other frequent corrections

  - Avoid using technical terms (*e.g.*, non-negligible) in normal contexts (⇒ "Significant" or "of low significance")

  - Both **"more metal-poor" and "metal-poorer"** are correct, but the second is better when it modifies a plural noun so is ambiguous; however, in many cases it is better to rephrase to "is richer/poorer in metal(s)".

  - The verb **suggest** takes the gerund not an infinitive, as do many others that mean different things depending on whether they are followed by a gerund or an infinitive ("Do you remember **seeing** this man before?" compared to "Remember **to put out** all the lights before you leave.")

  > Example: "Their article suggests doing the calculations again with the new parameters." The verb "suggest" is transitive, so the direct object is the verbal phrase. It can be rephrased as "Their article suggests that the calculations can be done again with new parameters."

Appendix A: References

Some of the available guides for scientific and technical writing that offer a wealth of examples and references along with clear explanations are

- Academic writing: Scientific Reports edited by Janet S. Dodd: `http://writing.wisc.edu/Handbook/ScienceReport.html`

- The ACS style guide: `http://www.oup.com/us/samplechapters/0841234620/?view=usa`

- The Online Technical Writing Textbook, which exists in hardback as *Power Tools for Technical Communication* `http://www.io.com/~hcexres/textbook`, by David A. McMurrey (Heinle Publishers, ISBN 0-15-506898-0).

Other useful sources are

1. the latest edition of *The Chicago Manual of Style* with their online search tools by subscription that give personal commentary by the editors on different points with lots of examples. `http://www.chicagomanualofstyle.org/tools.html`

2. Cambridge dictionaries on the web: `http://dictionary.cambridge.org`

3. Merriam-Webster at either `www3.merriam-webster.com/opendictionary/` or `http://www.websters-online-dictionary.org/`.

4. Online encyclopaedias such as (i) Wikipedia, especially its collection of pages under the headings of Natural Sciences, Mathematics, Technology, `http://en.wikipedia.org/wiki/Wikipedia:Browse` or (ii) Answers.com `http://www.answers.com/topic/accretion-disc` with "accretion disk" as the example here. This one uses several dictionaries and encyclopaedia articles, so may use several spellings of any words with a choice.

5. William Strunk & E.B. White *The Elements of Style* on "Bartleby.com", especially their sections of common usage errors at `http://www.bartleby.com/141/` or else useful general English guides from Hartford Community College/ Webster with a user-friendly portal at `http://grammar.ccc.commnet.edu/grammar`. Another resource could also be Fowler's *The King's English* also on Bartleby.com, but watch out for his irony and more complicated explanations compared to Strunk & White.

6. Mary K. McCaskill. "Grammar, Punctuation, and Capitalisation: A Handbook for Technical Writers and Editors". Langley Research Center (NASA SP-7084) at `http://www.sti.nasa.gov/sp7084/contents.html`.

Additional references on UK/US Spelling Conventions

If interested in less scientific differences, take a look at: Tripod.com's "US2UK"[8] or Krysstal's page of the differences[9]. Meanwhile, if you consider you are writing in another English dialect than either UK or US variants, then please let us know. If we cannot set our own spell checks on your file, however, then please include a list of what you consider the differences to be, which we can check if we feel it necessary: for instance, see Wikipedia's pages (gateway at `http://en.wikipedia.org/wiki/Wikipedia:Manual_of_Style#National_varieties_of_English`) on this concern, including the following statement on Australian English: "Both -ise and -ize are accepted, as in British English, but "-ise" is the preferred form in Australian English by a ratio of about 3:1 according to the Australian Corpus of English" (on `http://en.wikipedia.org/wiki/Australian_English`).

---

[8]`http://us2uk.tripod.com/dictionary.htm`
[9]`http://www.krysstal.com/ukandusa.html`

Scientific Writing for Young Astronomers
C. Sterken (ed)
EAS Publications Series, **49** *(2011) 107–134*
www.eas-journal.org

ASTRONOMY LIBRARIES – YOUR GATEWAY TO INFORMATION

U. Grothkopf[1]

Abstract. This chapter reviews services offered by astronomy libraries for assisting astronomers in their research. Special attention is given to the two most important information tools in astronomy: the NASA Astrophysics Data System (ADS) and the arXiv e-print server. Some less known features will be presented to lead to more efficient use of these tools. The core astronomy journals are explained, along with the open access concept and how it interacts with existing journals. The chapter also provides an introduction to bibliometric studies and shows why publication and citation statistics are important for both researchers and observatories. Finally, examples of cooperation between librarians and astronomers are presented in the context of working groups and conferences.

1   Introduction

The world of libraries is changing fast. The times when they were simply repositories of printed documents are long gone, because preserving the astronomy legacy and being curators of historic documents is important, but is not the only challenge librarians meet today. In the internet era, librarians deliver services directly to the scienstist's desktop, either by providing access to scientific journals and databases through assistance in using specific information tools or simply by answering reference questions via email or messaging systems. Astronomy libraries are a good example of this changing paradigm, because the vast majority of information resources are available electronically, and most astronomers are used to, and actually prefer, online resources over printed formats.

To use information efficiently, students and researchers need to acquire some basic skills. These can be divided into three parts.

- *Information literacy:* We have a surplus of information rather than a shortage. Scientists have to be able to locate the texts that are important for their research, otherwise they may waste time with information they do not actually need. Effective

[1] European Southern Observatory, Karl-Schwarzschild-Str. 2, 85748 Garching, Germany
e-mail: `esolib@eso.org`

DOI: 10.1051/eas/1149007

use of finding aids like library catalogs and reference works is called information literacy.

- *Computer literacy:* A certain understanding of the hardware, software, and network technology is needed to access information tools from the office, from home, or from abroad. This knowledge is called computer literacy.

- *Critical thinking skills:* Regardless of the medium (print or electronic, local or networked), no search result should be accepted without questioning its validity, correctness and completeness. Such critical thinking skills include common sense, as well as techniques for fully understanding a search result and putting it into context.

The combination of information literacy, computer literacy, and critical thinking skills is defined as information fluency, as explained in Viele (2008). Librarians will help develop these techniques and thus make the most of the time spent on literature and information searches.

2   What librarians can do for you

In today's interconnected world, a large fraction of the scholarly resources in astronomy are available online. Latest preprints are retrieved from the arXiv (astro-ph) e-print server, journal articles can be obtained via the NASA Astrophysics Data System (ADS) abstract service, and communication with colleagues of course mostly takes place by email. When asked what information services they use in their daily work, scientists often do not immediately think of their library. Yet, most astronomers do use libraries, even though they may not always realize it.

Start thinking of libraries not only as collections of printed books and journals, but also as gateways to electronic publications. Every time you download a recent journal article from the ADS, most probably you are allowed to do so because your library has paid the necessary subscription fee, negotiated access rights, and does the troubleshooting every time access is interrupted.

Librarians also compile web pages with link collections that provide access to electronic books, reference works, and databases. They help scientists discover new information retrieval systems and assist them in their research. To do this, librarians form professional networks to overcome the limited local resources and make use of joint collections, knowledge, and ideas provided by astronomy librarians worldwide.

In many organizations, librarians are in charge of institutional repositories. They hold introductory sessions to explain how to deposit manuscripts in the repository, make sure all necessary metadata get submitted, and are often experts in copyright and Open Access (OA). They also maintain databases of all papers published by the organization's scientists or, in the case of observatories, of those papers that use observational data from specific facilities. From these databases, statistics are derived and made available to management, governing bodies, and the public.

The library mission traditionally has been to fulfill the information needs of their users by selecting, collecting, preserving, and providing access to relevant resources. While many changes have occurred regarding the tools used in library management, this mission is still valid. In the digital age, personalized service, tailored to the needs of the library users, distinguishes libraries from software tools, see Grothkopf (2003).

Increasingly, research libraries act as meeting places. They provide space for both informal and formal meetings, ranging from the "science coffee" that takes place every morning to regular discussion groups and formally organized talks and presentations. Instead of the stereotypical quiet reading room, libraries often host collaborating scientists who appreciate the relaxed and inspiring atmosphere in the library.

Similar to the virtual "Web 2.0" concept, the traditional library has been replaced, or rather supplemented, by the "Library 2.0" that offers web pages with library blogs, RSS feeds, and interactive tools for contacting the librarians. Of course, especially in small, specialized libraries like those that often can be found in observatories, the most interactive way of consulting librarians is still to simply meet and talk to them. They are happy to provide research assistance, help you with finding quality information resources and suggest all kinds of tips and tricks you may not have heard of.

3 Two astronomy gorillas

Without doubt the majority of astronomers cover almost all their needs for scholarly information using only two resources: the NASA Astrophysics Data System (ADS) and the arXiv.org e-print server. ADS and arXiv clearly are the "gorillas" among their peers as the largest, most important, most visible resources. And they are both free of charge to the individual astronomer, hence available to everybody, no matter whether affiliated with an institution or not, or regardless how wealthy that institution is.

In the following, we briefly introduce both services and point out some features for using them efficiently.

3.1 NASA Astrophysics Data System (ADS)

The NASA ADS abstract service is operated by the Smithsonian Astrophysical Observatory (SAO) under a NASA grant. The main site is at `http://adsabs.harvard.edu`[1], with several mirror sites located around the world. It is the largest digital library in astronomy, providing links to more than 7 million records in astronomy, astrophysics, and physics, and it is used almost daily by the majority of astronomers. Access to subscription-based publications (including the two or three most recent years of the core astronomy journals) is granted only if the user or his/her institution subscribes to that publication. For the years 1975 through approximately 1995, the abstracts database contains data from a variety of sources; since 1995, most abstracts have been provided directly by publishers or journal editors. These references should be 100% complete. With less completeness depending on the subject area, but still at a high rate, ADS also provides information about older literature and, in many cases, offers links to scanned articles back to volume 1 of the most important astronomy journals. In addition to published literature,

---

[1] ADS main search screen: `http://adsabs.harvard.edu/abstract_service.html`

ADS provides information about preprints posted on arXiv.org. Preprints can be retrieved via the ADS the same day they become available on arXiv, so that there is no delay in finding the most recent postings.

ADS receives information about publications from a variety of sources, most notably directly from publishers on or even before the publication date, as well as from the arXiv preprint database. ADS users can query the database for immediate results or subscribe to the myADS service to receive regular notifications about new publications that correspond to search criteria set by the user.

3.1.1   Some search tips

The ADS main search screen is mostly self-explanatory. However, there are some features users do not necessarily know about. Some of them are explained below, and they have been available in the ADS since January 2009. Since the ADS is continuously being developed and improved, further capabilities can be expected in future.

First author searches: To limit searches to first authors alone, use ^ before the author's name. The resulting list will only include papers with that person as first author.

Filters: By scrolling down on the main search screen, you will reach the Filters section with a variety of options. For instance, to retrieve only publications from refereed journals, select "All refereed articles" in the first Filters section ("Select References From"). By using the "Select/deselect publications" option, users can enter journal codes of publications that are to be included in (or excluded from) the results. For a list of journal abbreviations, see the ADS Frequently Asked Questions page[2].

Citations: A very useful addition to the bibliographic record is the information about citations. To see the number of citing papers, go to the "Sorting" section towards the end of the screen and select "Sort by citation count" (or "Sort by normalized citation count" to normalize by the number of authors). Then enter your search. Alternatively, you can first enter your search and then use the "Sort options" pull-down menu in the upper right corner to sort by citations. On the results screen, you will see the number of citations above the title of each entry. It will be displayed with three decimal places (*e.g.*, 32.000) which are only important for the normalized citation count when citation fractions below 1 can occur.

For any given result set, you can get the list of citations for all or selected articles of the result list. The citation lists can include or exclude self-cites, and they can be limited to citing papers in refereed journals. In order to use these features, first enter a search on the main search screen. Make sure the whole set of results is displayed; if necessary, the number of items to be returned can be increased (default: 200). On the results page, select the records for which you want to see the citing articles, or scroll down below the last record and choose "Select All Records". Then scroll down further until you find the

---

[2]http://adsabs.harvard.edu/abs_doc/journals1.html

option "Get citation lists for selected articles" or "Get refereed citation lists for selected articles". To exclude self-citations, check this option.

A detailed description of the creation and use of citations in the ADS has been given in Accomazzi *et al.* (2007).

Export search results: Next to the Sorting section on the main ADS page, you will find the Format area, which governs the output format of search results. Instead of the default short list format, a variety of formats can be selected that allow easy importing into bibliographic tools like EndNote and ProCite. To add records to a Zotero library (see also Sect. 9 on organizing your own library), users can simply click on the paper icon that shows up in the URL field of the browser window. ADS records can also be displayed in a format ready for cutting and pasting into TeX manuscripts, *e.g.*, BIBTEX or AASTeX. Reference lists already formatted for specific journals can be generated, for instance for *Icarus, MNRAS,* and *Solar Physics.*

Historical literature: In addition to current publications, the ADS also provides access to historical literature. Together with the John G. Wolbach Library at the Harvard-Smithsonian Center for Astrophysics, the ADS digitizes microfilms from selected historical publications in astronomy. Although this compilation still lacks pagination so that one cannot go directly to a specific page, it still is very useful, in particular when looking for content that is otherwise difficult to find, such as observatory publications, reports, and annals. Lists of historical scans available so far are provided by the ADS[3].

Data links: The D links available for many records allow users to access the astronomical data used in these papers. This information is typically provided by the data center or observatory where the data are available. For instance, papers using data from telescopes of the European Southern Observatory (ESO) are linked by the ESO librarians to the programs that generated the data, and the D link for these records will take users directly to the ESO Archive where the data can be requested.

The "Select References In" option in the Filters area lets users limit queries to papers using data from a specific observatory, for instance HST, ESO, Gemini, Subaru, and Keck.

Catching errors: Some of the content available through the ADS is provided by libraries, even though the individual links do not reveal that information. In particular, librarians supply original bibliographic records, corrections to existing (especially historical) entries, and information about data links for their specific observatory.

3.1.2   Result lists: bibcodes, citation counts, and letter links

ADS result lists with brief entries display a variety of information above the authors and title. These are the bibcode of the record, the citation count, publication month and year, as well as several letter links, depending on availability.

---

[3] adsabs.harvard.edu/historical.html; adsabs.harvard.edu/journals_service.html

Bibcode: In astronomy, bibcodes are used to uniquely identify publications; they are assigned by the ADS. A bibcode consists of 19 characters. The first four represent the publication year, followed by up to five characters for the journal or book abbreviation, up to four digits for the volume, one optional character to indicate a special section (*e.g.*, Letters), up to four digits for the starting page, and finally the first letter of the family name of the first author. Unused characters/digits are filled with dots. A typical bibcode will be for instance 2005ApJ...635..260S, representing the article by Savaglio *et al.*, published in *ApJ*, vol. 635, starting on page 260.

Citation count: For the citation count of an article to be displayed, the respective selections in the Sorting section must be made, see above.

Publication month and year: The publication month and year are shown in the format MM/YYYY. This information is useful for instance when citations are compare, because it makes a difference how many months ago a paper was published.

Letter links: ADS entries provide links to additional information, symbolized by capital letters on results pages. Some of the information is stored on the ADS server, other links take users to services that reside elsewhere. The number and kind of links vary. The most important letter links are: A = abstract, C = citations (other articles citing the paper), D = data links, E = article fulltext in HTML, F = article fulltext in PDF, N = list of objects mentioned in the article from NED (NASA/IPAC Extragalactic Database[4]), O = associated articles (*e.g.*, errata), R = references (articles cited by a paper), S = list of objects mentioned in the article from SIMBAD database[5], X = links to articles in the arXiv database.

The letters are usually shown in blue. Fulltext links (PDF and HTML) of subscription-based articles will only allow access to the paper if you or your institution subscribe to the journal. However, E and F links of some articles will display in green, indicating that these articles are open access, *i.e.*, available to everybody. Whether open access is granted and how soon after publication, depends on the journal. For an explanation of open access, see Section 6.

3.2   *arXiv.org e-print archive*

The other large service in astronomy is the arXiv.org e-print repository[6]. The main server is located at the Cornell University Library. It provides open access to over half a million e-prints in physics, mathematics, computer science, quantitative biology, and statistics. For astronomers, the most important arXiv section is astro-ph, an e-print archive that dates back to April 1992. Like the ADS, also arXiv has mirror sites in various countries.

---

[4] http://nedwww.ipac.caltech.edu/
[5] http://simbad.u\discretionary-strasbg.fr/simbad
[6] http://arxiv.org

Originally developed by Paul Ginsparg when he was still affiliated with the Los Alamos National Laboratory (LANL), arXiv is now owned, operated, and mostly funded by Cornell University.

3.2.1   Manuscripts uploaded by authors

ArXiv/astro-ph is a collection of manuscripts posted by authors. There is no obligation for any author to make papers available, but the majority of astronomers post their papers on the e-print archive (Schwarz & Kennicutt 2004). The repository can be accessed via the web, or one can subscribe to email notifications and RSS feeds.

When submitting a paper, authors are asked to grant arXiv.org the non-exclusive right to distribute the article and certify that the work is either in the public domain or available under a Creative Commons (CC) license[7]. These licenses provide copyright holders with a mechanism for (partly or fully) waiving their rights in order to allow sharing of information, but CC license holders will still be credited if others use their works. The advantage is that obstacles typically arising from copyright laws can be avoided. Authors have to make certain that the license they grant to arXiv does not conflict with a journal agreement they may have signed[8].

To ensure that arXiv content is relevant to current research, an endorsement system was introduced in 2004. Since then, papers of first-time submitters have to be endorsed by existing members of the arXiv community before they are accepted by the e-print server. In January 2009, astro-ph was split into six sub-categories. Submissions have to be assigned to at least one of these subcategories. The split is meant to make announcements of new papers more manageable for those interested only in subsets of astro-ph[9].

As many authors post their papers before acceptance by a journal, several versions of the same paper can be posted while the refereeing process continues. The author decides whether every iteration is posted as a modified version. Typically, the text of the final posted version is identical to the published paper. The layout will most probably be different as the original typeset layout is copyrighted by the publisher, and most publishers will also require that you use and cite the refereed and published version of the paper, not simply the online arXiv version.

3.2.2   Finding your way through arXiv/astro-ph

On the main arXiv.org screen, astro-ph is listed as the first option in the physics section. Users can browse the entire database, go directly to entries made available today or during the recent week, or search astro-ph by author name, title, or abstract words and other categories.

For each entry, the arXiv number is shown, as well as title, author, and author-defined comments, *e.g.*, number of pages and figures, and status of the manuscript (submitted, accepted, revised version, etc.).

---

[7]http://creativecommons.org/license/
[8]http://arxiv.org/help/license
[9]http://arxiv.org/new/#jan2009

The arXiv papers should be cited in the format arXiv:YYMM.NNNNv# [category], for instance arXiv:0803.1234v1 [astro-ph], to indicate version 1 of paper number 1234, posted in March 2008 on the astro-ph server.

When citing manuscripts that have merely been submitted to journals and are not yet accepted, bear in mind that they are highly likely to undergo (sometimes substantial) changes before they are actually published. Indeed, scientists typically still cite the published version of papers whenever possible. Future changes may be one of the reasons, as well as the fact that the final reference reveals the journal where the paper will be published, so it conveys some of its reputation, *e.g.*, refereed journal versus conference proceedings. This effect is desired when compiling a publication list for the curriculum vitae or home page.

The vast majority of astronomers post their manuscripts on astro-ph; for instance, an estimated 80% of the papers we look at in the context of the ESO Telescope Bibliography (see Sect. 9.3) are submitted to the e-print server. Schwarz & Kennicutt (2004) find that, on average, *ApJ* papers posted on astro-ph are cited more than twice as often as those not posted. Does it matter where in the daily list of new postings a manuscript is positioned? Dietrich (2008a, 2008b) finds that papers appearing at or near the top of astro-ph listings receive significantly more citations than those on lower positions. Because submissions are listed in the order they arrive at the astro-ph server, with each new list starting with papers received at 4 pm (US Eastern time), authors do have some control over the rank of their articles. It remains to be seen whether Dietrich's findings are considered an unwanted drawback to the arXiv service by the managers at Cornell so that it leads to modifications in the organization of the ranking system.

4   Databases beyond ADS and astro-ph

With ADS and astro-ph being so dominant and relevant for a large number of queries, other services are often neglected, even though they may provide additional and sometimes unique information.

4.1   *SPIRES-HEP*

SPIRES-HEP[10] is the High-Energy Physics database of the Stanford Physics Information Retrieval System (later renamed Stanford *Public* Information Retrieval System). It was originally developed at the Stanford Linear Accelerator Center and is now a joint project of SLAC, DESY (Deutsches Elektronen-Synchrotron), and FNAL (Fermi National Accelerator Laboratory, Fermilab), managed by the SLAC library.

SPIRES-HEP provides bibliographic information about high-energy physics literature, including journal articles, theses, reports, conference papers, books, and others, with links to the fulltexts of these items. Among others, arXiv is one of the sources from where fulltexts of papers referenced in SPIRES-HEP can be obtained.

---

[10]http://www.slac.stanford.edu/spires/hep/

4.2 Google Scholar

Google Scholar (GS)[11] is a free research tool for scholarly literature and citations. While it is known that Google Scholar covers journal articles as well as conference proceedings, books, reports, and other media, and that even full-text repositories are searched, GS keeps users in the dark about exactly which sources it is using.

Similar to the main Google search engine, Google Scholar also does a very good job of correcting misspellings and suggesting different spellings including name variations. It covers a wide range of subjects, but seems to be generally stronger in the fields of social sciences, arts and humanities, and engineering (see for instance Harzing & van der Wal 2008). One drawback is that result lists can contain duplicates.

4.3 Scirus

Another literature search tool is Scirus[12]. It is maintained and provided free of charge by Elsevier Science, one of the largest scientific publishers. As of January 2009, over 480 million items are indexed, including refereed and unrefereed publications, repositories, science-related websites from the American Physical Society (APS), arXiv, the Institute of Physics Publishing (IOPP), NASA, and many other sources. Scirus offers a sophisticated advanced search interface that allows users to filter by date range, information type, and subject area (including astronomy). However, this tool only helps for retrieving publications and other information resources; citations are not provided.

4.4 ISI Web of Knowledge and Web of Science (WoS)

The ISI (Institute for Scientific Information) Web of Knowledge[13] is a platform that provides access to a variety of databases (*e.g.*, Inspec, Medline, Biosis) and tools for retrieving and analyzing research articles. Some backfiles, including citations, date back as far as 1900.

The citation database included in the ISI Web of Knowledge is called Web of Science (WoS)[14]. It covers over 9,000 publications in the sciences, social sciences, and arts and humanities, but is limited to journals, so that citations from books, reports, and other publications not included in the WoS are not counted. WoS is used by many organizations to analyze their scientists' research output, but it has been repeatedly criticized for omitting citations from books and for focusing mostly on English-language titles from the US and Europe.

In the past, the Journal Citation Report (JCR), which is prepared annually by the Web of Knowledge and which presents journal rankings, has been criticized for reporting flawed journal impact factors because of difficulties in handling possibly non-unique journal abbreviations. In astronomy, the *Astrophysical Journal* and *Astronomy &*

---

[11]http://scholar.google.com

[12]http://www.scirus.com

[13]http://www.isiwebofknowledge.com

[14]http://isiwebofknowledge.com/products_tools/multidisciplinary/webofscience/

Astrophysics have been particularly affected by this mistake (Sandqvist 2004). In the meantime, the error has been corrected and the acccurate journal impact factors published.

Unlike SPIRES, Google Scholar, and Scirus, WoS is not free. It is a commercial system, marketed by Thomson Reuters, that needs a subscription.

4.5　Scopus

Described as the "largest abstract and citation database of research literature and quality web sources" on their web site, Scopus[15] has become a serious competitor for WoS. As is the case with WoS, Scopus requires a subscription. The tool was developed and is marketed by science publisher Elsevier.

Scopus offers information from over 16,000 peer-reviewed journals, as well as conference proceedings, book series, scientific web pages, and repositories, including arXiv. Full-text articles can be accessed seamlessly, provided the necessary subscriptions are maintained. Scopus aims for complete coverage of their records from 1996 onwards, but historical material has also been added to the system, going back as far as 1823 in some cases. Citation analysis is provided from 1996 onwards.

One strength of Scopus are the innovative tools for analyzing publication and citation statistics, including the h-index and unique graphs.

The scope, coverage, and approach of Google Scholar, WoS, and Scopus are quite different, with WoS and Scopus providing substantial factual information, various search and ranking features, help pages, and guidance on how to use the services best. They act as an information hub (Jasco 2005). In contrast, Google Scholar applies the Google PageRank algorithm to display search results, which typically leads to the most popular hits being displayed at the top. Advanced search and display options are limited in comparison with WoS and Scopus.

Differences are becoming even more obvious when looking at citations. Meho & Yang (2007) find that the overlap in citations between Scopus and WoS is rather small, namely only 58%, and only 31% between Google Scholar and the union of WoS and Scopus. This emphasizes that none of these databases provides an absolute answer regarding completeness of search results and, specifically, numbers of citations. The three services complement each other, but none of them should be regarded as the perfect search system.

The main advice clearly is: when using publication and citation databases, never switch off your own thinking. Always evaluate all resources and put them in a context with other results.

5　Core astronomy journals

Every science has journals that are essential to that discipline. In astronomy, the number of these most important journals is small: only six titles are typically referred to as the

---

[15]http://www.scopus.com

"core journals." These are *Astronomy & Astrophysics (A&A), Astronomical Journal (AJ), Astrophysical Journal* and its *Supplement Series (ApJ/ApJS), Monthly Notices of the Royal Astronomical Society (MNRAS)*, and the *Publications of the Astronomical Society of the Pacific (PASP)*. These journals alone produced on average 36% of all refereed journal articles listed in the ADS for the years 2000–2007.

All core journals in astronomy are published on behalf of organizations or learned societies. *A&A* is published on behalf of the European Southern Observatory (ESO), *AJ* and *ApJ/ApJS* by the American Astronomical Society (AAS), the *Monthly Notices* on behalf of the Royal Astronomical Society (RAS), and *PASP* by the Astronomical Society of the Pacific (ASP). This means that, even if the journals are produced and marketed by commercial publishers, their (partly understandable) economical interests are somewhat leveled out by the societies who often influence or even control the subscription price.

The core astronomy journals are subscription-based; *i.e.*, the latest issues are only accessible with a valid subscription. However, all of them apply the "delayed open access" model. After two or three years, articles are made available to everybody (for a description of open access, see Sect. 6).

All of them are currently still available in print, as well as electronic format, even though some publishing models are already changing. For instance, the default subscription of the *ApJ Letters* includes electronic-only access as of 2009, and also the basic subscription price for *A&A* will soon only cover the electronic version; in both cases, the print version needs to be explicitly ordered (and paid for). These are first moves towards e-only publishing in astronomy, and it will only be a matter of time when the other core journals follow.

While librarians are not directly involved in the production of the journals, they pay the subscription fee, negotiate licenses (which define who can access the publication, and under which conditions), make journals available through web pages and library catalogs, and troubleshoot in case of access problems.

A&A, AJ, ApJ/S, MNRAS, and *PASP* are available through individual subscriptions that do not require customers to purchase a whole package of journals. Content from all titles can be found through the ADS or via astro-ph, provided that the authors submitted their manuscripts to the arXiv e-print server. In addition to these well-established finding tools, publishers are also starting to develop their own platforms to offer easy access to all their journals, plus additional information like commenting systems, readers' communities, etc. For instance, users of IOPscience[16] will be able to search latest and historical research articles published by IOP. Finding tools range from the established author/title/keyword searches to emerging technologies, such as tag clouds. Retrieved content also includes e-prints from arXiv (presented via an IOP-developed e-print interface) and certain web pages.

This trend away from static journals towards "virtual journals" or article databases can be expected to grow in future. It remains to be seen whether publishers' services will be able to compete with the "master search engine" provided by the ADS.

---

[16]http://iopscience.iop.org

6 Open access

6.1 *The OA concept*

A brief description of open access publishing is provided by Wikipedia: "publication of material in such a way that it is available to all potential users without financial or other barriers"[17].

The open access movement goes back to the 1960s, but has gained momentum mostly since the last decade of the past century. Among other reasons, it was prompted by the constantly rising subscription prices of scholarly journals that has led researchers, librarians, funding agencies, publishers. and others into discussion about how scientific publications can be made available in a better way. The main focus is on peer-reviewed, publicly funded literature. The idea is to remove some of the existing barriers, most notably price barriers (subscriptions, licensing fees, pay-per-view charges) and permission barriers (copyright and licensing restrictions, Suber 2007) and make them accessible to everybody. Of course "everybody" is a relative term, as some impediments continue to exist, for instance access limited to members affiliated with an institution, language barriers, accessibility barriers, etc.

There are various forms of open access:

- Self-archiving on authors' home pages: author self-archiving is often called the "green road" to OA. Most publishers (also commercial ones) grant their authors the right to make manuscripts, including the final version, available on the author's web page. If the final typeset version may be used, the publishers typically demand that the bibliographic details are given so that readers are pointed to the publisher's journal. The SHERPA RoMEO website[18] lists publishers' copyright and self-archiving policies.

- Self-archiving in a repository: these can be institutional or subject-based repositories. Many institutions have set up repositories where all affiliated researchers deposit their articles. Some scientific disciplines maintain central subject-based repositories. In astronomy, such a repository is the astro-ph e-print server.

  Self-archiving by scientists neglects two very important aspects: retrieval via a central access point, and preservation. Repositories are more reliable in this regard.

- Publishing in OA journals: typically called "open access publishing", this refers to journals that make their content available without subscription. In today's publishing landscape, many different flavors of OA are applied by publishers, ranging from delayed access (with free access to content after a given time) to institutional membership (whereby all papers are immediately available for those affiliated with the member institute), and on to the "author pays" model (*i.e.*, authors pay a certain fee to the journal, readers don't).

---

[17]http://en.wikipedia.org/wiki/Open_access_publishing, accessed Jan. 31, 2009
[18]http://www.sherpa.ac.uk/romeo.php

Even though removing barriers related to (sometimes unaffordably high) prices is one of the key incentives of the open access movement, it must be stated that serious OA advocates do not claim that the *production* of open access publications is without costs. The production of OA journals may overall be less expensive than conventionally published literature; however, it still demands funding that needs to be provided by someone. The word "free" (of charge) can therefore only refer to the readers of open access publications, not to the publication process at large.

One of the advantages open access provides for authors is better accessibility because, theoretically, every interested researcher can obtain the publications. Whether or not this actually leads to more citations is still being investigated. For a discussion of this topic in the case of *The Astrophysical Journal*, see Kurtz & Henneken (2007).

According to the key findings of the third ALPSP (Association of Learned and Professional Society Publishers) survey of Scholarly Publishing Practice[19], the proportion of publishers that offer optional open access (authors pay a fee) has grown from 9% in 2005 to 30% in 2008. However, the report concludes that "the take-up of the author pays open access option is exceedingly low." This statement is not surprising, given that (i) the fee authors have to pay for optional OA is sometimes as high as 3,000 Euros and (ii) that, especially in sciences like astronomy the majority of research literature is available either from the arXiv e-print server or directly from the journals through "delayed open access." As a result, astronomers may not be particularly attracted by a publishing model that requires further payments from them.

A description of the state of open access in astronomy can be found in Grothkopf & Erdmann (2008).

6.2   OA finding aids

Several finding aids can be recommended for retrieving open access publications.

6.2.1   Directory of Open Access Journals (DOAJ)

The idea of creating a Directory of Open Access Journals[20] goes back to the First Nordic Conference on Scholarly Communication held in Lund and Copenhagen in 2002[21]. The DOAJ Head Office is located at the Lund University Libraries. As of January 2009, the directory includes over 3,800 scientific and scholarly journals, of which more than 1,300 are searchable at article level. Open access journals are defined as peer-reviewed journals that "use a funding model that does not charge readers or their institutions for access" and that make all content freely available without delay.

---

[19]http://www.alpsp.org/ngen_public/article.asp?pfs=0\&did=47\&aid=27749\&oaid=-1
[20]http://www.doaj.org
[21]http://www.lub.lu.se/ncsc2002/

6.2.2   OAIster

OAIster[22] is a union catalog of digital resources, compiled by harvesting descriptive meta-data using the Open Archives Initiative Protocol for Metadata Harvesting (OAI-PMH). With a current staff of only three, the OAIster team has put together an impressive catalog of digital resources, including digitized books and articles, original ("born-digital") texts, audio files, images, movies, and datasets. These items are often located in the so-called deep web, hidden from search engines by firewalls. OAIster reveals these resources by using OAI-PMH and makes them available via a searchable interface. In December 2008, OAIster provided access to more than 19 million records from over 1,000 data providers.

6.2.3   ScientificCommons

The aim of the ScientificCommons project[23] is to "provide the most comprehensive and freely available access to scientific knowledge on the internet". Like OAIster, Scientific-Commons uses the OAI-PMH protocol to retrieve information from repositories. According to their website, ScientificCommons had indexed more than 13 million scientific publications as of January 2007. Author names are extracted across institutions and archives so that distributed publications can be found via one common interface. ScientificCommons also extracts professional relations between authors to make their research development transparent to the public. Personalization services include an option for authors to organize their publications directly with ScientificCommons.org in order to create their own researcher profiles.

ScientificCommons is a project of the University of St. Gallen (Switzerland), and is hosted and developed at the Institute for Media and Communications Management.

6.2.4   Directory of Open Access Repositories (OpenDOAR)

OpenDOAR[24] is a compilation of information about and links to academic open access repositories. It aims at increasing their visibility and retrievability by providing end-users with a search interface to locate specific archives by geographic location, subject, content type, and more. Physics and Astronomy are currently represented with 45 repositories.

From its start at the end of 2005, the size of OpenDOAR has increased quite steeply to reach its current coverage of approx. 1,300 repositories. Almost half of the included repositories are located in Europe, approx. 30% in North America, and around 20% in Asia, Australasia, and South America combined. OpenDOAR is maintained by the Unversity of Nottingham in the UK.

7   Problems with online documents

The internet and the advent of electronic publications have enormously modified how scientists use the literature. Especially in astronomy, we are used to finding (seemingly)

---

[22]http://www.oaister.org
[23]http://www.ScientificCommons.org
[24]http://www.opendoar.org

all research publications on the internet in electronic format. The ease of using online publications and their immediate availability should not let us forget that there are some severe obstacles on the way to the e-only scholarly society.

7.1 Long-term access

Ever since the advent of electronic publications, librarians have communicated the challenges in long-term access to the research community. Two major topics have always been of concern: archiving and preservation.

Archiving refers to questions such as:

- what material (scientific articles, electronic communication, databases, etc.) and which resources should be archived? (web pages only from certified providers? "everything"?)

- who should be in charge of archiving? (individual data providers? national libraries and archives? publishers?)

- where should the archived content be stored (private web pages, libraries, data centers, commercial archive providers, etc.) and how many mirror sites should exist?

Preservation of digital information resources encompasses assuring global perpetual access; provision of tools to read and interpret the data; migrating storage media to newer technologies, including all associated metadata, access rights, provenance information and, most importantly, the links within the document and those reaching out to other media items. Reference systems must be in place to make certain that the information can be found. The authenticity (identity) of electronic publications needs to be easily recognizable in order to determine which version of the document is being used, and their links and connections with other digital objects must be transferred to each newer storage medium. Finally, mechanisms are needed to guarantee the integrity (intactness) of digital information.

7.2 Completeness of digitized material

In order to make historical literature, *i.e.*, material originally published on paper, available online, numerous scanning projects are underway at observatories, universities, the ADS, and other data providers. How complete are the results? At the minimum, all projects scan the scientific content, *i.e.*, the articles published in the journals, magazines, conference proceedings, books, or other items that are being digitized. But what happens to other parts, such as errata, news sections, advertisements, letters to the editor, front and back matter, obituaries? Depending on the treatment received during and after digitization, such material may be lost or destroyed.

7.3 Deleted items

In the print environment, the path of publications is quite well-defined. After production at the publisher, it is sent to libraries where the material is displayed and later bound into volumes and placed on shelves, available for future access whenever needed. Nothing gets lost or is changed, the scholarly content is archived the way it was originally published.

In the digital environment, things are more short-lived. For instance, what happens if a journal publisher finds out that an article plagiarized someone else's work, or if the content, if acted upon, poses a serious health risk? In some cases, it might be considered necessary to remove the article from the publisher's website, even though it had been formally published (see for instance the Elsevier policy on article withdrawal[25]).

Other examples are Wikipedia pages of ceased projects or comments and contributions posted to mailing lists and blogs that may be removed. While such items are intrinsically more ephemeral, they show that the availability of online content should not be taken for granted.

7.4   Broken links

In today's use of web resources, it often seems to be sufficient to indicate specific pages through Uniform Resource Locators (URLs). While many of them reveal which organization or company hosts the source (*e.g.*, en.wikipdia.org) or what specific subdirectories are on that server (*e.g.*, .../wiki/Url), they can also simply consist of figures and then not tell users anything (at least not at first sight) about their origin and host.

The real danger, however, is that URLs point to one exact location. Should the location change or any part of the string be modified, users will not be able to retrieve the intended web page, but will be sent to an error page.

To avoid such broken links, a better naming system has been developed: Digital Object Identifiers (DOIs[26]). DOIs are names, not locations, and provide a unique, permanent digital identifier to objects. Mostly used for scientific publications, it has been widely applied by publishers. By the end of 2008, approx. 40 million DOI names had been assigned.

A DOI follows a defined structure and may look like this: 10.1065/abc123defg, where 10.1065 is the prefix. The first two digits are the directory code; 10 indicates the publishing sector. 1065 represents a given publisher; abc123defg is the suffix. It is chosen by the publisher and has to be unique within the realm of the prefix. DOIs can be resolved through browsers, for instance at the website dx.doi.org. The site name and the DOI are seperated by a forward slash[27].

The system is managed by a membership consortium called the International DOI Foundation. The official link registration agency for scholarly and professional publications is CrossRef[28], a citation-linking network that covers millions of articles and other content items. In order to have DOIs assigned to their publications, publishers have to become CrossRef members and pay an annual fee.

---

[25]http://www.elsevier.com/wps/find/intro.cws_home/Article\Withdrawal
[26]http://www.doi.org
[27]for instance, http://dx.doi.org/10.1065/abc123defg
[28]http://www.crossref.org/

8 Everything is on the internet ... really?

Compared to other sciences, astronomy traditionally has been on the forefront of the move towards electronic publications. In 1997, only two years after the start of electronic publishing, almost half of the peer-reviewed literature in astronomy was available in electronic format (Boyce 1998). This may lead to believing that by now, *everything* has been made accessible online. This assumption is understandable, but wrong. Information resources that are not available on the internet often are forgotten, neglected, ignored.

Information not easy to find includes material such as:

- special supplements accompanying (in particular historic) publications, such as photographic plates;

- literature in radio astronomy; this refers in particular to historical publications, but also to recent literature published in engineering journals;

- observatory publications, especially those older than two or three years;

- books and conference proceedings, in particular if published more than approx. five years ago;

- non-English language literature.

When looking for literature, it is crucial to use various retrieval tools and not rely on "quick & dirty" internet searches that may tempt you to believe that the result is already complete, while in reality it is not.

9 Organizing a personal library

Having access to the literature is one part of using information, but keeping your own library of papers that are of interest may be referred to in your own publications is another. An additional difficulty arises from journals typically applying their own style for citations. This means that authors need to follow certain guidelines when they compile reference lists for their manuscripts. These guidelines are different for each journal.

For many years, various systems have been around to help scientists with this task. These were mostly commercial programs that had to be purchased and installed on a computer, so-called reference management software. Such software stores information about (scientific) articles, and the bibliographic citations can then be output in various formats to be used in reference lists.

More recently, open source software systems have been made available that are available to scientists at no cost or for a small fee. The majority of these programs are still used locally on the user's desktop, but some have a web complement or are even entirely web-based.

Standard features that most reference management systems fulfill include:

- create, import, export records;
- assign notes, keywords, or tags;
- create collections for specific topics or purposes;
- store copies of the articles (in PDF format) for off-line reading.

Of special interest here is in which format records can be exported. Astronomers usually compile reference lists in one of the following two ways.

- LaTeX: in many cases the list of references is the last section of a manuscript. References are manually added as soon as they are cited in the text. The LaTeX reference list is directly integrated in the manuscript

- BibTex: a BibTex reference list (.bib file) requires a bibliography style file outside of the manuscript which calls the reference that is cited in the manuscript.

Many reference management systems are able to handle BibTex references (*e.g.*, export records in BibTex format), while only a few have a direct link (export feature) for LaTeX references. If references need to be exported from systems without direct LaTeX integration, it is recommended to export them first to a "plain" export format (*e.g.*, a clipboard) and copy and paste them from there into the LaTeX reference list. A good comparison of reference management software can be found in Wikipedia[29].

In the following, we briefly explain four reference management systems and outline some specific characteristics. These four systems are JabRef, Papers, Zotero, and Mendeley.

- **JabRef:** The JabRef[30] reference manager is available as freeware under the GNU General Public License. It is platform-independent so that it runs equally well on Windows, Linux, and Mac OS. JabRef's main features include the ability to

  - import, organize, annotate, and archive

  records. Existing BibTex libraries (.bib files) can be imported, and the developers claim that it is also possible to fetch entries from the arXiv eprint website. PDFs can be read offline so that users are independent of internet access. Exporting records requires custom export filters, *i.e.*, layout files that reside outside of JabRef. These must be written by users and can be shared via SourceForge.net. JabRef is built around BibTex and is a good tool for users who are comfortable using BibTex.

  Users can create various collections, for instance, to group records by topic or to keep all records pertaining to specific reference lists in one place. In addition, individual records can be highlighted for easier retrieval.

---

[29]http://en.wikipedia.org/wiki/Comparison_of_reference_management_software
[30]http://jabref.sourceforge.net/

- **Papers:** Papers[31] is a reference management software that has been developed specifically for the Apple Macintosh operating system. It is not free of charge; a single-user license costs EUR 29 as of mid-2009. Papers is integrated with the iPhone so that articles can be synchronized.

The software allows users to

  – find, import, organize, annotate, and archive

libraries. As is the case with JabRef, import plug-ins exist for BibTex files and other formats. A specialty of Papers are several built-in search engines that provide direct access to articles databases, including the ADS and arXiv. Without having to leave Papers, users can search the repositories, save repeating searches, and use some of their search features, like limiting them to first author searches or other specific fields. Retrieved records can be imported with one keystroke. The search history is accessible from within Papers so that searches can be re-used easily.

Papers intends to encourage users to read articles on-screen by showing PDFs in full-screen mode, providing the opportunity to annotate them, and by remembering the last reading position in the file. In addition to bibliographic information about articles, Papers also allows users to store email addresses of author names and passwords for access to fulltexts of journals.

Exporters exist for BibTex, Word2008, and CSV (comma-separated values) formats. Unfortunately, the system does not provide an option to add keywords to records or tag them, which is quite surprising for a software with otherwise high functionality.

- **Zotero:** Zotero[32] is a plug-in for the FIrefox 3.0 (or higher) browser. It can be used on Windows, Mac, or Linux machines. The main features include options to

  – find, download, organize, tag, and archive

records. Existing libraries can be imported from intermediate formats, *e.g.*, BibTex. The main specialty of Zotero, however, is the ease with which records can be captured directly from the web. Simply by clicking on the little notepad or paper icon in the URL-field, web page content can be imported and then be processed further, *e.g.* by assigning records to collections that have been created or by attaching files (PDFs, screenshots, etc.) to records. Related papers can be linked to each other.

Libraries (collections) can be exported for instance to BibTex or Rich Text Format. Alternatively, a bibliography of selected records can be created using any of the journal styles included in the large style gallery. The second option will provide records formatted according to that particular journal's citation style. Unfortunately, no astronomy-related journals are currently included in the list of journals, so that astronomers will probably simply end up creating a bibliography of the

---

[31]http://mekentosj.com/papers/
[32]http://www.zotero.org

records they want to export, copy it to the so-called clipboard, and then format it manually.

The Zotero window can be collapsed entirely in the Firefox browser, in which case only the Zotero icon is visible in the lower right corner. When in use, the program will either occupy the lower third of the browser window or it can be opened to fill the whole window. Zotero is organized in three columns, with the left one holding the various collections, the middle one titles, authors, and publication years of the individual papers, and the right column showing details of the currently edited record. Article PDFs can be stored with the records for offline use, and users can assign tags and notes to records. In addition to published literature, Zotero can also take snapshots of web pages, which can then be viewed offline, annotated, and marked using a highlighter tool or little post-it stickers.

In order to visualize records in the collections, Zotero offers an option of viewing a timeline. Timelines are organized on three levels; filters can be assigned to limit the number of records shown, and highlighting specific keywords allows users to view publishing trends.

- **Mendeley:** Mendeley[33] is a reference management tool for the desktop (Windows, Mac, Linux) with an online counterpart. The program is free of charge, but users need to sign up in order to download the software. The current version (as of mid-2010) is still beta (0.9), indicating that Mendeley is still being developed. The program intends to provide a system with which users can

  – manage, share, and discover

research papers. Records can be imported directly from the ADS, arXiv, Google Scholar, and other databases. Exporting records from Mendeley is currently possible in Word and OpenOffice Writer, using more than 100 citation styles. Once again, astronomy journal styles are not available.

A special feature of Mendeley is its ability to automatically extract information from PDFs through drag and drop. From files dropped into Mendeley, the system will import bibliographic details along with DOIs, arXiv IDs, and other information. PDFs can be annotated using a highlighter tool or sticky notes. Mendeley's powerful full-text search capability retrieves search terms not only in author names, titles, and keywords, but in the entire text of articles.

The program is very much geared towards establishing collaborations among colleagues and tries to act as a social networking tool. By using the web part of the system, users can form networks and invite others to share their papers and discover new ones. Research trends are visualized, for instance, by displaying the most popular authors and topics and through statistics about the papers in the collection.

---

[33]http://www.mendeley.com

All systems described above have some advantages and some disadvantages. JabRef is platform-independent, but its use is not very intuitive, because the program is built around BibTex and therefore will be used mostly by those who are familiar with BibTex. The second program, Papers is more user-friendly, and it searches the ADS and arXiv, but is only available for the Mac and is not available entirely free of charge. Zotero, the Firefox plug-in, conveniently captures metadata directly from the web, thus allowing easy import of records of articles and web pages, but the style gallery for exporting records is not very suitable for astronomers so that manual formatting of references is necessary. Mendeley provides a remarkable full-text search capability aimed at building communities among scientists. Right now, the functionality is still somewhat limited, but the developers seem to be moving fast in order to incorporate more features.

In the end, the decision to use a particular reference management system will depend on personal preferences and on which system matches most of a user's selection criteria.

10   Bibliometric studies

Publicly funded organizations typically have to report back to their management and funding agencies about how their resources were spent and whether they provide the expected output. For scientific entities, the output is increasingly measured by bibliometric methods. Often librarians are in charge of these studies.

Bibliometrics can be defined as a set of methods used for publication and citation analysis to explore the impact in the respective field. Observatories tend to use such metrics to evaluate the research output based on data from their telescopes and instruments, and the acceptance of these facilities among the user community.

Some commonly used bibliometric methods are listed in Table 1. All of them have both advantages and disadvantages. For instance, counting the number of publications reveals the productivity of an entity, but gives no information about the impact and importance of these publications. Citation analysis, assuming that important articles are cited more often by other authors, does reflect the impact among peers. However, citation counts can be inflated, for instance by erroneously counting citations of different scientists who have the same name and initials, because of friends and colleagues who systematically cite each other, or even because of articles that only point out incorrect findings in the cited work (Meho 2007). Calculating the mean or median number of citations per paper has the advantage of allowing comparison (bearing in mind the difficulties intrinsic to such comparisons, see Sect. 9.1) of publications from different scientists or from various facilities, but high citations to only a few articles can increase the average value, which may wrongly suggest that the majority of papers included in the result set have received high cites.

Another measure are so-called High-Impact Papers (HIPs), a term coined originally by ISI (now Thomson Reuters) and later applied to publications in astronomy (Meylan *et al.* 2004). An HIP is defined as a paper that belongs to the 200 highest cited refereed papers in a given year. Once this subset is defined, *e.g.*, by using the ADS Abstract Service, papers using observational data are analyzed for the observing facilities that provided the data. This method shows interesting trends, in particular if applied over several years.

Table 1. Advantages and disadvantages of metrics.

	Good	Bad
# Publications	Productivity	No impact
# Citations	Impact	Can be inflated
Mean/median cites per paper	Allow comparison of different eras	Reward low productivity
"High-Impact Papers"	Show trends	Favor "hot topics"
h-index	Productivity + impact	Determined by years of operation

However, it is very time-intensive to compile and may favor "hot topics" that receive a lot of momentary attention.

A few years ago, Hirsch (2005) proposed a new method, the so-called *h*-index. It aims at quantifying a researcher's scientific research output in a more balanced way by combining numbers of publications and numbers of citations, hence productivity and impact. The idea seems to be simple and yet more effective than many other metrics: the *h*-index is defined as the number of publications that have received at least *h* citations. Using the ADS or other citation databases (*e.g.*, Web of Science, Scopus), the value can be computed quite easily. First, submit a query to retrieve all papers for which you would like to calculate the *h*-index. For instance, such a query can be for all publications by a specific author, using data from a particular telescope, or all scientists affiliated with a specific organization. The set of results needs to be ordered by decreasing number of citations, so that the paper with the highest number of citations resides at the top of the list, the one with the lowest citation number at the bottom. At that point, *h* can be found where the rank number of the paper (starting with rank number 1 at the top) is higher than or equal to the number of citations the paper has received.

As might be suspected, *h* computed as described cannot be the same for authors that have been publishing for many years and for those who just entered research; naturally, the senior researcher will have published many more articles that will have gathered many citations in the course of the years. Similarly, *h* will also not be comparable across disciplines or even across subfields of disciplines; *e.g.*, papers published in the subject area of planetary astronomy will on average have different *h*-indices than, for instance, UV astronomy publications. To make comparisons possible, at least to some extent, Hirsch introduced a second measure, the so-called *m*-parameter, which is calculated by dividing *h* by the number of years. For instance, this can be the number of years of scientific activity of researchers or the number of years of operation of facilities: $m = h/t$.

10.1 Caveats of citation analysis

Regardless of which bibliometric method is chosen, there are some general problems one should keep in mind.

- *Incompleteness:* by now, several web-based services have been developed that provide citation information. Some of them (*e.g.*, Web of Science, Scopus) require a subscription, others (*e.g.*, ADS Abstract Service, Google Scholar) are free-of-charge. The sources and technologies the services use to obtain and analyze the data vary, and citation statistics can be vastly different. Therefore, it cannot be emphasized enough that one single citation database will never give the full picture.

- *Incorrectness:* the sources of incorrect citation information can be manifold. A simple, but frequent, one is that authors make mistakes in their lists of references. Accordingly, these references cannot be resolved by the citation databases and are not assigned to the intended cited paper, *i.e.*, the citation will be ignored. Another reason for errors lies in different abbreviations used for the same journal. References to each of them may then be assigned to seemingly different journals, and citations may be counted in parallel rather than cumulatively.

- *Citing behavior:* it should be remembered that many authors prefer to cite well-established scientists, those that are cited by everybody. There may be many reasons for this behavior: it may arise from a wish to "go with the flow" of what other authors do; it may save time to use reference lists from papers written previously; authors may have a wish to show that they read the "classics" in the field, or perhaps they want to show "solidarity" with friends and close colleagues. Regardless of the exact motivation, this creates a bias towards highly cited scientists and puts young and less connected authors at a disadvantage.

- *Multi-author papers:* authors who publish in large collaborations[34] by definition create a professional network of many colleagues who will (presumably) cite this paper in their future publications. Obviously, this leads to a much larger basis of potential quotes than papers published by single authors or small groups of co-authors will have. A way to handle references to such many-authored papers would be to assign only normalized (proportional) citation credits, *i.e.*, 1 divided by the number of authors.

One more word of caution. It is very bad style to quote without actually having read the article, even though everyone in the field seems to quote it. One also would risk attributing something to the cited author that he or she did not actually write.

10.2   Alternative measures

Citation analysis is an important part of bibliometrics, but it has one severe disadvantage: citations do not measure actual use: *i.e.*, how often were papers consulted with or without subsequent inclusion in the final list of references?

A new measure to better reflect the use of publications has been introduced by the ADS. Instead of citations, readership information is being collected. A *read* is counted as each time a user requests information via one of the ADS letter links. In 2002, 50%

---

[34]In a recent paper, Crabtree (2008) found that the average number of authors has increased from 2.5 in 1980 to almost 7 in 2006. In the context of the ESO Telescope Bibliography, we noticed a 2008 Nature paper with no fewer than 93 authors.

of the requested items were abstracts, followed by one of the fulltext versions (38%) and citation lists (8%) (Kurtz *et al.* 2003). Using their database of logs, the ADS team can identify relations between the private act of reading a paper and the public one of actually citing it. The reads history is available for ADS entries from the record's abstract page.

A similar approach is pursued by Citebase[35]. This experimental citation database looks at *downloads* of papers from arXiv.org. As of January 2009, only the UK-site is monitored, but the other arXiv mirror sites are ignored, including the main server in the U.S. This means that the number of downloads per e-print displayed is notoriously too low. It will be interesting to follow the development of this project.

On a more general level, access and download statistics can also help for evaluating the use of journals and their overall acceptance within the community of scientists. Libraries occasionally consult such statistics to determine which journal subscription should be continued and which may no longer be necessary.

10.3   Telescope bibliographies

At many observatories, librarians are in charge of maintaining databases of papers written by staff or other astronomers that use data from the observatory's facilities. These databases are typically compiled by scanning the important astronomical literature (manually or semi-automatically), trying to locate references to instruments and telescopes that were used to gather the data.

There are many reasons for maintaining telescope bibliographies, for instance, to measure the scientific impact of telescopes and instruments. The performance of existing facilities can also provide important guidelines for future facilities. Funding authorities often request bibliometric studies because they help them to understand the overall productivity and impact of the funded organization.

Following the observatory's policies for acknowledging telescope use is also important for the authors. Some funding agencies demand that their support be acknowledged, and also many observatories ask (or even oblige) authors to mention use of the facilities in a footnote[36]. Mentioning facilities in the footnote will also help compilers of telescope bibliographies to keep track of the scientific papers published with their data and may influence observing proposal committees and other groups for granting future observing time. Last, not least, publishing with data from renowned facilities is also advantageous for authors as it increases their visibility and helps them to be recognized by colleagues.

A list of telescope bibliographies from major ground-based and space-based observatories is available from the ESO library[37]. In addition, the ADS provides access to several keyword lists, among them many observatories. On the main ADS search screen, scroll down to the Filters section and choose "Select References In / All of the following groups" and select the observatory of interest. Note, however, that the selection criteria, the years covered, and the range of completeness will differ. To find out the exact coverage, click

---

[35]http://www.citebase.org
[36]For instance, the policy for publishing with ESO data can be found at
http://www.eso.org/sci/observing/policies/publications.html
[37]http://www.eso.org/libraries/publicationlists.html

on the facility's name, which will take you to a short explanation. As explained before, the D link of ADS records will lead to the data center or archive from where the data used in the respective paper can be retrieved.

11   Cooperations of librarians and astronomers

Traditionally, librarians have undertaken the role of mediator between information providers and scientists. Many librarians regularly inform the astronomers at their institute about new developments from publishers and database hosts, and report back any concerns, questions, as well as suggestions for improvements. You will know the librarian at your home institute, and if you are looking for a librarian at another institution, the Directory of Astronomy Librarians and Libraries[38] may be of use.

On a larger scale, there are some more formal ways to enhance communication between the various groups, in particular, an IAU Working Group and the LISA conferences.

11.1   IAU Commission 5: Working Group Libraries (WG Lib)

The Working Group Libraries is associated with the IAU commission 5 "Documentation and Astronomical Data." It aims at better cooperation between astronomers and librarians. The working group was officially recognized for the first time in 1990, but librarians were involved in Commission 5 activities even before that. Information about past and present activities can be found in the *Reports on Astronomy* (published every three years in the *Transactions of the IAU*), as well as on the working group's web pages[39].

During IAU General Assemblies, the Working Group often hosts so-called business meetings. These sessions are open to everybody interested and typically focus on data access and preservation, science metrics, and the role of librarians in fostering research.

11.2   Library and Information Services in Astronomy (LISA)

In 1988, the first conference on Library and Information Services in Astronomy was held in Washington, DC, at the U.S. Naval Observatory. Since then, LISA has been developed into a series of conferences, and four further meetings took place, each attended by more than 100 participants. LISA conferences provide a unique meeting point for librarians, astronomers, publishers, and computer specialists from around the world. The wide geographic distribution of participants is indeed one of its main characteristics.

Thanks to the relatively small number of core journals, in many cases coupled with quite generous funding, astronomy libraries have often had access to evolving technologies earlier than other subject areas, and have applied them early on in their day-to-day work. In addition to changing technology, the professional role of librarians has undergone considerable development. This is reflected in the topics covered at conferences. Today, LISA is a forum to exchange experiences, view topics from different angles, and gather information about emerging fields of interest in astronomy libraries.

---

[38]http://www.eso.org/libraries/addresses/addresses.html
[39]http://www.eso.org/libraries/IAU-WGLib/

A description of the history of LISA meetings, their logistics, and topics can be found in Corbin & Grothkopf (2006); a web page provides links to LISA conferences, as well as proceedings of past meetings[40].

12 Conclusions

The publication paradigm continues to shift from printed material to electronic formats for scientific literature, leading to an evolved concept of library services and information access. Librarians are taking on new and diversified roles. Observatory librarians fulfill the information needs of astronomers and engineers and develop digital services to give greater access to scientific content. They support their library users when conducting research and provide help so that existing tools can be used efficiently.

In this chapter, we have introduced the varied research assistance libraries provide. Some tips and tricks were provided to help especially young astronomers to retrieve information from the ADS digital library and the arXiv (astro-ph) e-print repository. An overview was given of the most important astronomy journals and the concept of open access (OA) publishing. Because of the widespread habit of astronomers of posting manuscripts on astro-ph, OA affects astronomy and astrophysics less than many other subject areas, yet it should not be assumed that nothing remains to be done.

Scientists have gotten very used to the idea that all necessary information is available (typically free-of-charge) on the internet. However, there are some severe problems regarding long-term access, completeness of digitized material, and broken links, to name just a few. Librarians and information providers are working towards more reliable systems, but it is essential that scientists are also aware of these problems.

Increasingly, librarians maintain telescope bibliographies, *i.e.*, databases of articles that use observational data. From these databases, publication and citation statistics are derived to provide information about the scientific output of specific facilities to management and funding organizations. Telescope bibliographies conclude the "life cycle" of observations because they close the loop from observing proposals to observations and to the published results, and from there back to the underlying data.

Finally, we have shown examples of close cooperation between librarians and astronomers that increase communication and, ultimately, lead to more efficient research.

This research has made extensive use of the NASA Astrophysics Data System; it is a pleasure to thank the ADS team for their always fast and friendly service and inspiring exchange of ideas. My sincere thanks also go to Christopher Erdmann, ESO Library, for reviewing the manuscript and helpful discussions.

References

Accomazzi, A., *et al.*, 2007, Creation and use of citations in the ADS, in: *"Library and Information Services in Astronomy V"*, ed. S. Ricketts, C. Birdie, E. Isaksson, San Francisco, CA, Astronomical Society of the Pacific, ASP Conf. Ser. 377, 93 [arXiv:cs/0610011v1] [cs.DL], http://arxiv.org/pdf/cs/0610011v1

---

[40]http://www.eso.org/libraries/list.html

Boyce, P.B., 1998, Electronic publishing in astronomy, in: *"The impact of electronic publishing on the academic community"*, ed. I. Butterworth (London, Portland Press)
www.portlandpress.com/pp/books/online/tiepac/session1/ch3.htm

Corbin, B.G., & Grothkopf, U., 2006, LISA – the Library and Information Services in Astronomy conferences, in: *"Organizations and Strategies in Astronomy"*, vol. 7 ed. A. Heck (Kluwer Academic Publishers, Dordrecht), 285
http://www.eso.org/libraries/articles/lisaconferences.pdf

Crabtree, D.R., 2008, Scientific productivity and impact of large telescopes, in: *"Observatory operations: strategies, processes, and systems II"*, ed. R.J. Brissenden & D.R. Silva, SPIE Conf. Proc. 7016, pp. 70161A-70161A-10

Dietrich, J.P., 2008a, The importance of being first: position dependent citation rates on [arXiv:astro-ph], PASP, 120, 224
http://www.journals.uchicago.edu/doi/pdf/10.1086/527522

Dietrich, J.P., 2008b, Disentangling visibility and self-promotion bias in the [arXiv:astro-ph] positional citation effect, PASP, 120, 801
http://www.journals.uchicago.edu/doi/pdf/10.1086/589836

Grothkopf, U., 2003, From books to bytes: changes in the ESO Libraries over the past decade, The ESO Messenger, No. 112 (June 2003), 51
http://www.eso.org/sci/libraries/articles/books2bytes.pdf

Grothkopf, U., & Erdmann, C., 2008, Open access – state of the art, IAU Information Bulletin, 102, 64, http://www.eso.org/libraries/IAU-WGLib/IAU_IB102p64.pdf

Harzing, A.W.K., & van der Wal, R., 2008, Google Scholar: the democratization of citation analysis? Ethics in Science and Environmental Politics, 8, 62
http://www.harzing.com/download/gsdemo.pdf

Hirsch, J.E., 2005, An index to quantify an individual's scientific research output, Proc. Natl. Acad. Sci. USA, 102, 16, 569 [arXiv:physics/0508025v5] [physics.soc-ph]
http://arxiv.org/pdf/physics/0508025v5

Jasco, P., 2005, As we may search – comparison of major features of the *Web of Science, Scopus,* and *Google Scholar* citation-based and citation-enhanced databases, Current Science, 89, 1537,
http://www.ias.ac.in/currsci/nov102005/1537.pdf

Kurtz, M.J., Eichhorn, G., Accomazzi, A., *et al.*, 2003, The NASA Astrophysics Data System: obsolescence of reads and cites, in: *"Library and Information Services in Astronomy IV (LISA IV)"*, ed. B.G. Corbin, E.P. Bryson & M. Wolf, U.S. Naval Observatory, Washington, DC, 223, http://www.eso.org/sci/libraries/lisa4/Kurtz.pdf

Kurtz, M.J., & Henneken, E.A., 2007, Open access does not increase citations for research articles from The Astrophysical Journal [arXiv:0709.0896v1] [cs.DL]
http://arxiv.org/pdf/0709.0896v1

Mahoney, T.J., 2007, Correspondence to the Editors of "The Observatory", Obs., 127, 401, reprinted: IAU Information Bulletin, 2008, 102, 70
http://www.eso.org/sci/libraries/IAU-WGLib/IAU_IB102p70.pdf

Meho, L., 2007, The rise and rise of citation analysis, Physics World, January 2007, 32
http://eprints.rclis.org/archive/00008340/01/PhysicsWorld.pdf

Meho, L., & Yang, K., 2007, A new era in citation and bibliometric analyses: Web of Science, Scopus, and Google Scholar, subm. to Journal of the American Society for Information Science and Technology, 58 [arXiv:cs/0612132v1] [cs.DL]
http://arxiv.org/pdf/cs/0612132v1

Meylan, G., Madrid, J.P., & Macchetto, D., 2004, HST science metrics, PASP, 116, 790,
    http://www.journals.uchicago.edu/doi/pdf/10.1086/423227

Sandqvist, A., 2004, The A&A experience with impact factors, in: *"Organizations and Strategies
    in Astronomy"*, vol. 5. ed. A. Heck (Kluwer Academic Publishers, Dordrecht), 197. Preprint at
    http://www.astro.su.se/~aage/OSA5.pdf

Schwarz, G.J., & Kennicutt, R.C., 2004, Demographic and citation trends in astrophysical journal
    papers and preprints, BAAS, 36, 1654 [arXiv:astro-ph/0411275v1]
    http://arxiv.org/pdf/astro-ph/0411275v1

Suber, P., 2007, Open access overview: focusing on open access to peer-reviewed research articles
    and their preprints,
    http://www.earlham.edu/~peters/fos/overview.htm

Viele, P., 2008, Information fluency and physics curriculum: faculty/librarian collaboration, APS
    Forum on Education Newsletter, Summer 2008
    http://www.aps.org/units/fed/newsletters/summer2008/viele.cfm

Scientific Writing for Young Astronomers
C. Sterken (ed)
EAS Publications Series, **49** *(2011) 135–157*
www.eas-journal.org

FROM YOUR PAPER TO VIZIER AND SIMBAD

L. Cambrésy[1], F. Genova[1], M. Wenger[1], C. Loup[1], F. Ochsenbein[1]
and T. Boch[1]

Abstract. Scientific results strongly rely on previous studies, experiments, and observations. A huge quantity of data is produced by astronomers and needs to be available to the whole community. To support astronomers in their daily research work, the *Centre de Données astronomiques de Strasbourg* (Strasbourg astronomical data center – CDS) collects, verifies, homogenizes, and organizes information – in particular, those published in academic journals – in the most appropriate and comprehensible way. This CDS goal can reach the highest level of quality only through a close collaboration with journals, authors, and referees. This paper presents the succession of processes leading published data to the CDS databases, focusing on the strategy that maintains a high level of quality. Authors and referees are strongly encouraged to actively contribute to this endeavor. A few examples of how CDS databases are used through CDS services are also presented.

1 Introduction

The *Centre de Données astronomiques de Strasbourg* (Strasbourg astronomical data center – CDS) was created in 1972 by the French agency in charge of ground-based astronomy, which is now called INSU (*Institut National des Sciences de l'Univers*), as a joint venture with Strasbourg Louis Pasteur University. Louis Pasteur University merged on January 1, 2009, with the two other local universities to form the *Université de Strasbourg*. The CDS was created as the *Centre de Données Stellaires* (Stellar Data Center), but in 1983 it was decided to extend its scope and to consider all astronomical objects beyond the Solar System.

The high-level goal of CDS is to *"collect, homogenize, preserve and distribute astronomical information for the usage of the whole astronomy community"*. Its initial charter is still relevant:

- collect "useful" data on astronomical objects in electronic form

---

[1] Observatoire Astronomique de Strasbourg, France

- improve them by critical evaluation and combination

- distribute the results to the international community

- conduct research using these data.

From the beginning, CDS has gathered two main types of information:

- astronomical catalogs, with a catalog service shared with several other data centers around the world, which was extended to include tables published in journals. These catalogs were distributed on demand on magnetic tapes and then through an ftp service. The VizieR service, available since 1996, has added a browsing capability (Ochsenbein *et al.* 2000).

- object cross-identification and bibliography were first made available through two different data bases, the *Catalogue of Stellar Identifications* (CSI – Jung & Bischoff 1971; Jung & Ochsenbein 1971), which aimed at cross-identifying a few fundamental star catalogs, and the *Bibliographic Star Index* (BSI – Cayrel *et al.* 1974). SIMBAD (Set of Identifications, Measurements, and Bibliography for Astronomical Data, Wenger *et al.* 2000) was created in 1981 from the merging of the BSI and of the CSI.

The *Dictionary of Nomenclature of Celestial Bodies outside the Solar System* (Lortet *et al.* 1994) and the Aladin interactive star atlas (Bonnarel *et al.* 2000) are two other important CDS products which will be referred to in the following.

Involvement of active researchers in the CDS staff has always been recognized as an important factor for success, since it is the only way to ensure the required scientific expertise on the service content and to make sure that the services are suited to the constantly evolving needs of the scientific community. To maintain a high level of scientific activities, the CDS was created inside and as a part of an Observatory, which provides an active research environment. The CDS team integrates astronomers, computer engineers who manage the databases and the user interfaces, and highly specialized bibliographers, who are called *documentalistes* in French – this shows that the CDS librarians are specialized in extracting "information" from and building metadata for documents. Information and metadata of interest for CDS are not only the usual bibliographic information managed in libraries (authors, publication references, etc.), which constitute the relatively easy part of the work. The point here, as will be shown in the following, is to check the description of the catalogs and tables stored in VizieR or to extract important data about astronomical objects for SIMBAD.

In the following, CDS services and procedures are described with the aim of showing how they are built and how the scientific community can help the data center in its constant quest for quality, along with a few hints to how they can be used. Only a selected set of information about the CDS services is highlighted in this paper. For more details, the reader is referred to Genova *et al.* (2000) and to the other papers of the series published in the April 2000 Special Issue of *Astronomy & Astrophysics Supplement Series* (Vol. 143, No. 1, several of which have been cited above): "The CDS and NASA ADS resources: New tools for astronomical research", and for a global more recent view of CDS history and evolution, to Genova (2007).

2 The VizieR database

2.1 Content

The link between VizieR and SIMBAD is not always clear to new users since the two databases are indeed very different. SIMBAD results from the cross-identification of major catalogs, tables and individual objects published in the literature. VizieR comes from a different approach: the astronomical catalogs are kept in their original form, in what we call the *FTP archive* (a set of files and directories openly accessible from cdsarc.u-strasbg.fr). In this repository, one catalog occupies one directory and is always accompanied by a *standard description*, in the form of a file named ReadMe described in the next Section 2.2. These homogeneous descriptions are the key to the VizieR system, a relational database gathering the tabular data (catalogs) existing in the FTP archive; in addition, VizieR gives access to very large surveys that are not part of the FTP archive because of their huge size.

VizieR currently provides access to the most complete library of published astronomical catalogs and data tables available online (more than 8000 catalogs in June 2010), organized in a self-documented database. A brief summary of VizieR is given below (Sect. 2.3); the contents of VizieR are illustrated by Figure 1, which represents the spatial distribution of $\sim 10^9$ sources available in VizieR after removal of the all-sky surveys.

Fig. 1. VizieR surface density map in galactic coordinates, representing $\sim 10^9$ sources after removal of all-sky surveys USNO-B1 ($\sim 10^9$ sources) or 2MASS. The horizontal bar in the central region is due to the GLIMPSE catalog (map built on June 2010 with more than 7500 catalogs).

2.2 The ReadMe file

Information that describes the data – the metadata – is traditionally presented in the introduction of the printed catalog, or explained in one or several published papers presenting and/or analyzing the cataloged data. Metadata play a fundamental role. First they provide scientists with information about the context of the data so that they can make their judgment about the suitability for their project. Also a minimal knowledge of the metadata is required by data processing systems in order to merge or compare data from different origins – for instance, the comparison of data expressed in different units requires a unit-to-unit conversion, which can be performed automatically only if the units are specified unambiguously.

This need for a description that is readable both by a computer and by a scientist led to a standardized way of documenting astronomical catalogs and tables, promoted by the CDS and adopted by other data centers in the form of a dedicated ReadMe file associated to each catalog (Ochsenbein 1994). The ReadMe description file (see example Fig. 2) starts with a header specifying the basic references of the catalog – title, authors, references – and contains a few key sections introduced by standard titles like *Description* or *Byte-by-byte Description of file*. Such a file is relatively easy to produce by someone who knows the catalog contents – or better by the authors themselves.

The most important part of the ReadMe file is the *Byte-by-byte Description*, which details the table structures in terms of formats, units, column naming, or labels (some conventions are used for label assignments for consistency), existence of data (possibility of unspecified or null values), and brief explanations.

This standardized way of presenting the metadata proved to be extremely useful, especially for data-checking and format conversion. Many errors were detected in old catalogs simply because a general checking mechanism became available.

Astronomy & Astrophysics authors are expected to supply the documentation of their data in this simple form. Template files, as well as a few tips on how to create the ReadMe file, are accessible on the VizieR webpage[1]. The ReadMe files and the data files are then checked by a specialist who contacts the authors if errors are detected or when changes are necessary to increase the clarity or homogeneity of the description.

The data accessible in the *FTP Archive* are not restricted to tabular data: if historically astronomical catalogs have been the first type of data distributed on electronic media by the CDS in the early seventies, the astronomical data available from this repository now include also spectra, images, data cubes, etc. However, data on the public repositories are in principle restricted to *scientific data*, *i.e.*, data that can be re-used in the data processing tools used in astronomy. This restriction means that FITS images, for instance, can be found (FITS includes essential details about origins, sky location, etc.), but JPEG or PDF images are generally excluded.

2.3 VizieR

Once a catalog, as described according to the above conventions, is part of the *FTP archive*, it becomes visible to any interested scientist. Search engines provided by the CDS help to locate these catalogs from their textual contents with *GOOGLE*-like queries. Textual searches, however, cannot answer to questions like *"find in all tables of this*

---

[1]http://vizier.u-strasbg.fr/doc/submit.htx

```
J/A+A/466/137      JHKs photometry in Westerlund 2          (Ascenso+, 2007)
================================================================================
Near-infrared imaging of Galactic massive clusters: Westerlund 2.
    Ascenso J., Alves J., Beletsky Y., Lago M.T.V.T.
    <Astron. Astrophys. 466, 137 (2007)>
    =2007A&A...466..137A
================================================================================
ADC_Keywords: H II regions ; Clusters, open ; Photometry, infrared
Keywords: open clusters and associations: individual: Westerlund 2 -
          stars: pre-main sequence - infrared: stars

Description:
    Photometry for the stars detected with SOFI, NTT in J, H and Ks in the
    field containing the massive cluster Westerlund 2 (also know as RCW 49)
    associated with the HII region Gum29. For each star detected in all
    three bands the table contains the ID, equatorial coordinates, J, H
    and Ks magnitude and photometric errors. For the stars detected only
    in Ks or in H and Ks the magnitude and photometric error fields in the
    absent bands are filled with the values 99.99 and 9.999 respectively.

File Summary:
--------------------------------------------------------------------------------
 FileName   Lrecl  Records   Explanations
--------------------------------------------------------------------------------
ReadMe        80       .     This file
w2phot.dat    65     5727    Positions and magnitudes of stars in Westerlund 2
--------------------------------------------------------------------------------

Byte-by-byte Description of file: w2phot.dat
--------------------------------------------------------------------------------
  Bytes Format Units   Label    Explanations
--------------------------------------------------------------------------------
   1-  4  I4    ---     Seq      Identification number
   6-  7  I2    h       RAh      Right ascension (J2000.0)
   9- 10  I2    min     RAm      Right ascension (J2000.0)
  12- 16  F5.2  s       RAs      Right ascension (J2000.0)
     18   A1    ---     DE-      Declination sign (J2000.0)
  19- 20  I2    deg     DEd      Declination (J2000.0)
  22- 23  I2    arcmin  DEm      Declination (J2000.0)
  25- 29  F5.2  arcsec  DEs      Declination (J2000.0)
  31- 35  F5.2  mag     Jmag     ?=99.99 Magnitude in the J band (1)
  37- 41  F5.3  mag     e_Jmag   ?=9.999 Photometric error in the J band (1)
  43- 47  F5.2  mag     Hmag     ?=99.99 Magnitude in the H band (1)
  49- 53  F5.3  mag     e_Hmag   ?=9.999 Photometric error in the h band (1)
  55- 59  F5.2  mag     Ksmag    ?=99.99 Magnitude in the Ks band (1)
  61- 65  F5.3  mag     e_Ksmag  ?=9.999 Photometric error in the Ks band (1)
--------------------------------------------------------------------------------
Note (1): The values 99.99 for the magnitude and 9.999 for the errors
     correspond to sources for which we do not have photometry in the
     corresponding band.
--------------------------------------------------------------------------------

Acknowledgments:
    Joana Ascenso, joanasba(at)astro.up.pt
================================================================================
(End)     Joana Ascenso [CAUP, Portugal], Patricia Vannier [CDS]   25-Jan-2007
```

Fig. 2. Example of ReadMe file for catalog I/221.

archive which contains entries related to point sources in a 2-arcmin region around my target". Such queries have to deal with numeric values and have to know what a position is and which coordinate system is used in each table, etc.

These queries are possible with VizieR, the database created from the contents of the *FTP archive*. VizieR is basically a relational database that keeps both the original data (each catalog or table of the FTP archive becomes a table of the database) and the associated metadata extracted from the `ReadMe` files (stored in a set of dedicated tables). These metadata play the essential role of giving a meaning to all the numbers and symbols included in the catalogs and tables. VizieR structure permits queries to be constrained by any parameter, *i.e.* column, of the tables. Some further metadata homogenization is performed to better interpret the parameters contained in the thousands of tables of VizieR, essentially by assigning a *Unified Content Descriptor* (UCD) to each column of each table. The UCD, originally defined at the CDS (Ortiz *et al.* 1999) is shorthand to qualify the meaning of a parameter in a thesaurus of the existing parameters used in astronomy (*e.g.* positions, magnitudes, oscillator strength), which is now developed in the framework of the *Virtual Observatory*.

Besides the homogenized metadata, VizieR also introduces *links* to associated data. As an example, when a paper contains images or spectra associated to observed or modeled objects, VizieR may provide a link to the images or spectra related to the corresponding object entry in the table. Such links may also point to other VizieR tables, to other databases (SIMBAD, NED, HyperLeda), or to repositories (SDSS, HST archive, etc.).

VizieR also gives access to large surveys, such as 2MASS or USNO-B1. The access to these huge surveys, which is not possible in the *FTP Archive* because of the volume of data, is optimized to give fast answers to positional queries.

Currently (June 2010), VizieR contains more than 8000 catalogs made of 19 000 different tables for a cumulative number of 3×10^5 columns, *i.e.*, an average of 2.3 tables per catalog and 12.5 columns per table. VizieR responds to an average of 1.2×10^5 queries/day.

2.4   Usage of VizieR

Access to VizieR by Web is available by script, or by programs. The data published in VizieR also become accessible by the Virtual Observatory, which means that the tabular material, once properly documented and ingested into VizieR, becomes directly accessible by any tool aware of these standards. As a practical example, tables describing astronomical sources become visible in Aladin (Sect. 5). The results available this way are also easily retrieved for the creation of large samples required to derive new astronomical results.

An accurate description of the data related to astronomical objects is also an important step before integrating them into SIMBAD (see Sect. 3.3); therefore, do not miss the documentation of your data!

3   How your data are integrated in SIMBAD

3.1   Acronym creation in SIMBAD

Astronomical object designation, either called name or identifier, is a key point for SIMBAD since it refers unambiguously to an object. Some rules need to be defined for clarity.

The general syntax of an identifier is the abbreviated catalog name, or acronym, followed by one or several fields, which can be numbers, coordinates, strings, etc. An identifier is case and space insensitive.

When no general catalog name can be used, the acronym is directly based on the author's initials between square brackets with the year of publication of the paper. For example, HD 97300 is also named [FL2004] 44 after Feigelson & Lawson (2004).

Object names, such as Vega and Altair, but also Barnard's star, Crab Nebula, Sgr A, HDFN, or HDFS, are stored in the database in a specific catalog called "NAME", while star names in constellations, such as Lyra, are stored in the catalog "*" (*e.g.* * gam Lyr), and variable stars (such as RR Lyrae) in the catalog "var" (also called "V*"). To some extent, identifiers can be shortened (*i.e.* NAME can be omitted), or can be typed according to some accepted usage. A set of rules, based on regular expressions and substitutions, try to normalize the typed identifier before querying the database.

Greek letters should be abbreviated as three letters: alf, bet, for α and β, but also mu., nu., and pi. (with a dot), for μ, ν, and π. This is required to avoid ambiguities in some identifiers. Constellation names should be abbreviated with the usual three letters: alf Boo, del Sct, FG Sge, NOVA Her 1991. The full list is available online.

Identifiers of a multiple system may generate a list of all objects belonging to the system. For instance, ADS 5423 calls for the four components, A to D, of the stellar system around Sirius. This is only true for some specific identifiers.

Clusters that have no NGC or IC number are named under the generic appellation Cl followed by the cluster name and number: *e.g.*, Cl Blanco 1 is the 1st stellar cluster named by Blanco. Stars in clusters follow complex *historical* rules. They may belong to a *main* designation list or to subsequent lists. NGC 5272 692 is star 692 in the list by Von Zeipel, considered as the main list. Subsequent lists have designations starting with Cl* such as Cl* NGC 5272 AC 968 (list by Aurière & Cordoni), Cl* Melotte 25 VA 13 (13th star in the list by van Altena for Melotte 25 – the Hyades cluster).

3.2   Integration of objects cited in the text of papers into SIMBAD

As a general rule, all objects cited in a paper are stored in SIMBAD with a reference to this paper and relevant data if any. This has long been done by *documentalistes* reading journals and checking them for object citations. Additional paths for object entry have been made possible by the development of electronic journals. In particular, the CDS and *Astronomy & Astrophysics* collaborate to provide links between object names in papers and SIMBAD, by asking authors to tag *important* object names. These tags are used to identify objects to be referred to in SIMBAD (Sect. 3.2.1). Most of these objects, however, are entered through a systematic scan of the papers or by a semi-automated processing of published tables. The procedure used for naming objects and cross-identifying them is described in Sections. 3.1 and 3.2.3.

3.2.1   Tagging astronomical objects

The *Astronomy & Astrophysics* instructions ask authors to *"surround any astronomical object in your text, as well as in short tables with the command"*:

```
\object{<objectname>}
```

Objects tagged with this macro will appear linked to the corresponding SIMBAD page in the journal electronic edition. Only the authors are allowed to tag the objects. About 20% of the references in *Astronomy & Astrophysics* contain tagged objects, which explains why many astronomical objects do not have a SIMBAD link in the electronic edition.

The list of publications that cite an object is provided for each entry in SIMBAD. Although the tag does not require having a paper included in the list of references associated with an object, it significantly shortens the delay of this process. Tagging is therefore suitable for rapidly informing the community interested in an object that a specific paper is relevant.

Tags seem an efficient tool for allowing authors to directly interact with the SIMBAD content, however reality is not that simple. Naming an object is indeed a critical step that can often lead to ambiguities. Each tag requires a check by the SIMBAD team at the time of the paper's electronic publication, because it can be a new name, which is not in SIMBAD yet, or because of a potential careless use of nomenclature by the author **and** a lack of attention from the referee. Regarding the authors' part, CDS provides a web service to perform all the verifications by submitting the *.tex or the *.obj files (the .obj is generated by the compilation of the LaTeX file):

```
http://vizier.u-strasbg.fr/viz-bin/Object
http://vizier.u-strasbg.fr/viz-bin/Sesame.
```

The *Astronomy & Astrophysics* instructions to authors provide these two addresses and *"encourage you to test the astronomical objects using the sites and easy tools available at the CDS"*. Journals from the *American Astronomical Society* have adopted similar methods for maintaining links from the electronic publications to SIMBAD. Despite these tools, the number of necessary actions to build a working link is increasing. In 2002, 13% of the papers had tags and 25% of them needed *corrections*. In 2006, 25% of the papers had tags and 45% of them needed *corrections*. *Corrections* mean incorrect designations or creation of new acronyms for SIMBAD (new objects). The latter case is obviously not an author's mistake. The increase in corrections is a consequence of tagging the objects present in long tables. Since these objects, only cited in long tables and not explained in the text, are not individually important for the paper, authors are now asked to tag only the most relevant objects, meaning no more than 10–20 tags per paper. Long tables must be excluded systematically. In the present paper, tags are not used at all because none of the cited objects are studied, and this reference is obviously not relevant to any science topic.

Besides this, correct designations are sometimes ambiguous because they may refer to several different possible objects, for instance, the result of a SIMBAD query for *Perseus* yields 17 possible entries, including a molecular cloud, a star association, a superbubble, a spiral arm, a Seyfert 2 galaxy, or even a galaxy cluster. It is therefore critical to avoid ambiguous names. As shown in Figure 3, *Perseus* alone will point the Perseus OB 1 star association, so more specific names must be chosen to reach the correct target, hence the importance of checking links before submitting the paper.

object	OType	J2000 position	Nrefs	Resolver
perseus	Assoc*	02:21:00.00 +57:36:00.0	77	Simbad: Ass Per OB 1-
perseus molecular cloud	MolCld	03:36:00.00 +31:00:00.0	101	Simbad: NAME PER CLOUD
perseus cluster	ClG	03:19:48.00 +41:30:42.0	791	Simbad: ACO 426

Fig. 3. Summary of objects tagged in a test manuscript with the \object macro. This is obtained with the tool available at the URL http://vizier.u-strasbg.fr/viz-bin/Object.

3.2.2   Designations of astronomical objects

SIMBAD is not limited to tagged objects. All papers from ~30 journals are read by the CDS staff to extract all astronomical objects cited in the text. Taking the example of *Perseus* again, staff have to decide which of the possible objects the paper actually deals with. The first step is to recognize an astronomical name in the text, given the prolific imagination of some authors. One might naively think that any word with special characters or numbers is the signature of an astronomical object, but this is not necessarily true. A field designation for instance will probably have the typical structure of an object name although its inclusion in SIMBAD is not desired. Also some object type names are based on their prototype, such as T Tauri stars or FU Orionis stars. When the star name is used as object type, the star T Tauri or FU Ori is not to be added to the list of objects for the reference, and, symmetrically, the reference should not be added to the star T Tauri or FU Ori. The context of the paper is usually clear enough on this aspect for a well-trained *documentaliste* to find out.

Confusing names such as LW1 are more difficult to track. LW1 is an ISOCAM filter onboard the ISO satellite, and it is also a globular cluster named NGC 1466. LW1 stands for Lynga & Westerlund (1963) source number 1 and comes from a table of 483 sources with a record number as a first column. If this paper were being ingested at the CDS now, the LW1 name would be [LW63] 1. Using a prefix formed by the author's initials with the year of publication within square brackets has the advantage of linking the designation to a publication and prevents any confusion.

For example, if one decides to name his/her object M1, SIMBAD will remove the confusion with the famous Messier 1, the Crab nebula, by using the prefix. Unfortunately, some authors might cite the object but drop the prefix, and it will become more and more difficult to know which M1 it is, except maybe for the specialists of the relevant topic. Confusion will spread over the literature. Another example presented in Figure 4 shows the content of the dictionary of nomenclature[2] for the identifier L. The second column gives the actual designation used in SIMBAD. Lynds dark nebulae should be called LDN followed by a number, such as LDN 113, for instance. The name encountered is often L113 instead. This happens so often that the ambiguity is actually solved by

---

[2]Dictionary link on http://cdsweb.u-strasbg.fr/

Acronym	Use Format	Year	1st Author	Obj. Type
L	L FFFF-NNNNA	1942	LUYTEN W.J.	PM*
(L)	[AEL81] {L}NNa	1981	ALLOIN D.+	Part of G
(L)	[H68] {L}	1991	WAKKER B.P.+	HVC
(L)	[L63] NNN	1963	LINDSAY E.M.	Em*
(L)	[L65] N	1965	LYNGA G.	HII
(L)	HHMMSS.s+DDMMSS [L92] {Sgr D} N {Sgr E} NN {SA} FFF LN {TPhe} A	1992	LISZT H.S.	(Rad)
(L)	[L92b] {Mark} AN {RU} NNN A {PG} HHMM+DDd A	1992	LANDOLT A.U.	*
(L)	Lal NNNNN	1950	BAILY F.	*
(L)	LAWD NN	1949	LUYTEN W.J.	WD
(L)	LBN LLL.ll+BB.bb	1965	LYNDS B.T.	HII
(L)	LDN NNNNA/NNNNWW	1962	LYNDS B.T.	Dark N
(L)	[LGC91] {ACO} NNNN NNNN	1991	LUCEY J.R.+	G + * in ClG
(L)	Lin NNNa	1961	LINDSAY E.M.	Em. Obj.
(L)	Lindsay NNN	1958	LINDSAY E.M.	OCl
(L)	LLNS NNNNN	1976	LODEN L.+	*
(L)	Loden NNNN	1977	LODEN L.O.	Loose and,or poor OCl
(L)	[PWA92] NN	1992	PRUSTI T.+	*
NLTT	L FFFF-NNNNA	-	LUYTEN W.J.	PM*
OL	Ohio NNN.N/-NNN.N	1967	SCHEER D.J.+	(Rad)

Note: Usage of acronyms in parentheses like 'L' should be *avoided*

Fig. 4. Dictionary of nomenclature for the acronym L.

SIMBAD where L113 is correctly interpreted as LDN 113. Nevertheless, if one writes L113, thinking to the open cluster Lindsay 113, the result will be wrong.

On the other hand, the square-bracket prefix is not suitable for *official* catalogs. It would be meaningless to add a prefix to SDSS or 2MASS sources since the designation is formatted, so unique, and comes from a released catalog. There cannot be any confusion. Actually there are, because some authors decide to alter the official name by truncating the correct designation or by changing one digit (when the format is based on celestial coordinates). This should be absolutely forbidden. Official names must be respected even if based on coordinates, regardless of the slightly different coordinates found by follow-up observations.

Finally, papers would gain in clarity if astronomical objects were systematically accompanied by coordinates or with a second known designation in case of cross-matches. This simple recommendation would avoid a lot of ambiguities in the object identification.

3.2.3   Cross-identifications proposed in papers

Among the recent progress in astronomy, multi-wavelength analysis is certainly one of the most spectacular. This approach has been supported by SIMBAD since its beginning in the parent database CSI. One of the main SIMBAD goals is to ingest the cross-identifications found in the literature. The scientific impact is considerable for characterizing of astronomical objects and for understanding the underlying physics.

Cross-identifications will eventually lead to spectral energy distributions. Because of their crucial role, cross-identifications are introduced as carefully as possible in SIMBAD.

Unfortunately, the literature cannot always be trusted. A nearest neighbor may be a reliable cross-identification if wavelengths, sensitivities, and resolutions are similar. If not, this technique is not sufficient so a scientific analysis is required. For instance, the cross-identification of a 2MASS source (arcsec resolution) with an IRAS source (arcmin resolution) can be very hazardous, except if the object type, say a carbon star, ensures that a single 2MASS object can explain the far-infrared flux measured by IRAS. Also the cross-identification of optical with radio or X-ray sources can be risky if the method is limited to the nearest-neighbor criterion. Since authors are specialists in their domains, we cannot reach the equivalent level of expertise in all fields of astronomy to decide whether a cross-identification proposed in a paper can be trusted, but we try to ignore hazardous matches as much as possible.

Incomplete cross-identifications are often found in papers. A typical author's choice, for instance, is to provide a list of sources in the optical and to add the SDSS magnitudes without providing the SDSS names. There are definitely cross-matches, but they are implicit and we might have trouble finding out why. Ambiguities (*i.e.* two nearby objects) generally occur, implying a serious loss, sometimes waste, of time for the SIMBAD team when a lot a papers are waiting to be ingested, and loss of useful information known to the authors.

3.3   Integration of catalogs and tables in SIMBAD

While short published tables are integrated in SIMBAD manually by the bibliographers in charge of those articles, longer tables listing more than about 50 objects are integrated in a semi-automated way by other specialized bibliographers. Such tables are first archived electronically in the VizieR database. These VizieR electronic tables are then used and processed to include the objects and new data in SIMBAD in the most consistent way. In the first step, cross-identifications with objects already in SIMBAD are systematically searched, using coordinates and names, and checking the consistency between object types. Most of the work of the bibliographers consists in adjusting the search parameters to optimize the cross-identifications, and in checking ambiguous cases individually, with the support of Aladin facilities and of the expertise of CDS astronomers. Once all problems have been solved, the integration of the tables in SIMBAD is done automatically. Although such a procedure can be time consuming, it is one the major goals of SIMBAD to perform careful cross-identifications between various catalogs. In case of doubt, the conservative rule is to create a new object rather than to risk an erroneous cross-identification.

In past years the number of tables published in the articles has dramatically and continuously increased. Among all the tables archived in VizieR, about one third does not have to be integrated into SIMBAD, because they are not relevant (*e.g.* list of atomic transition probabilities) or because it would not improve or would even degrade the level of information: lists of detailed measurements, lines or variability studies for instance; photometric CCD surveys without characterization of the sources in very crowded fields (like in globular clusters); extremely huge catalogs like 2MASS or the SDSS. Pieces of huge catalogs are integrated via shorter tables of follow-up studies, or sometimes by dedicated operations, to cross-identify the content of SIMBAD with new large catalogs. Still, since 2005 the CDS has faced a total of 400–450 tables to be integrated into SIMBAD, with sizes ranging from a few 10 to 10^5 objects.

The global CDS strategy is to take full advantage of the complementarity between VizieR and SIMBAD: all tables are made available in VizieR, and a large fraction of them in SIMBAD – in priority those for which the SIMBAD added value is the most prominent.

To avoid too much delay in the integration of more important tables or catalogs, the strategy is that the CDS team, astronomers and bibliographers, perform a scientific expertise of all tables. The fundamental data and measurements to be entered in SIMBAD, as well as the main parameters to search for cross-identifications, are setup by the team, as is an order or priority. Then tables are classified into 3 groups: group 1 (first priority, 45%), to be integrated in SIMBAD as soon as possible; group 2 (second priority, 45%), to be entered in SIMBAD when possible; group 3 (10%), lists of known objects without new data, which can be done entirely automatically. Various criteria can yield a first-rank priority. In particular, the level of information input, which includes accurate redshift or velocity measurements, memberships, expert cross-identifications especially between different wavelength ranges, and improvement of data (coordinates, spectral types, magnitudes, etc.). The priority also goes to outstanding fields of research, such as cosmological deep fields, brown dwarfs, X- or gamma-ray researches, and identification of YSOs. The number of citations (relative to the date and journal of publication) and number of catalog queries in VizieR are checked as well.

The content of the published tables is fundamental to gaining or wasting time and information. For instance, a table listing SDSS sources can be integrated into SIMBAD two to three times faster if the authors use the latest SDSS version and list the correct and entire name. Careful cross-identifications, expertized by the authors, are also very valuable, because they save time by including the table and improve the level of information. The content of the SIMBAD database is made not only by the CDS team, but also by the astronomers themselves through the information included in their papers and the tables therein.

Information from tables is thus stored in VizieR and/or SIMBAD. To help users retrieve the information available at the CDS more easily, the team is now working on a general portal able to access all CDS resources, including of course VizieR and SIMBAD, from a single query.

3.4   Hierarchical links

In SIMBAD, some tentative hierarchical links between astronomical objects were first introduced through object types, such as a star or galaxy in a cluster or group. This, however, has turned out to be inadequate, since it did not allow all cases of hierarchy to be included and was often reflecting more an observing method than a real link. In particular, CCDs surveys of clusters easily lead to a field of *stars in cluster* distributed according to the instrument field of view rather than the actual cluster shape. Finally, the link by object type did not allow keeping track of either the reference in SIMBAD or the parent body, if relevant. A new tool was thus developed to include hierarchical links between SIMBAD objects. The link provides the following information: status (*parent* or *child*), probability of membership (given in the reference or estimated by SIMBAD), and bibcode of the reference providing the link. Hierarchical links can be physical, like objects belonging either to a cluster or a galaxy, multiple systems, or to a shell and or central star of a supernova. They can also be non-physical due to the improvement of instruments and resolution. For instance, an IRAS source can be resolved in several components in MSX; then, the IRAS source is the *parent* of the MSX sources, although it is no longer a single astronomical object but historic artifact worth keeping track of.

Thus, when a link is indicated on a SIMBAD object, it does not mean that the two objects are actually physically linked: a link is also put for foreground or background sources with a low or no membership in order to save this information of non-membership which can be as important as membership for some studies. In some cases, one astronomical object may have several parents. For instance, one candidate Wolf-Rayet (WR) star in a galaxy was later resolved into one B star and one WR star; the parents of both stars are the historical WR candidate and the galaxy. Hierarchical links have been systematically entered in SIMBAD as a test since 2008 for tables and catalogs archived in VizieR (Sect. 3.3). No hierarchical links are put for short lists of objects entered manually by the bibliographers, as it would always require the expertise of an astronomer. We do not consider performing a full update of the SIMBAD database for hierarchical links, which would be incredibly time-consuming. However, users can send a request to CDS[3] proposing to include them for a dedicated catalog or areas, and providing us with as much information as possible.

4 The SIMBAD database

This section focuses on the user's point of view by describing the content of SIMBAD to illustrate the final result of the data entry process described in the previous sections. The query modes are described in the Appendix.

4.1 Content

SIMBAD is continuously updated and contains 4.8 million objects in June 2010, with several kinds of data.

- **Identifiers**: census of the different names found in the literature to designate an astronomical object. The list relies on successive cross-identifications. The total number of identifiers in SIMBAD reaches 13.9 million.

- **Basic data**: mostly data that allow the identification of the object, *i.e.* coordinates, proper motions, parallaxes, radial velocity and redshift, spectral type, morphological type, galaxy dimensions, and magnitudes.

---

[3] mail to: question@simbad.u-strasbg.fr

The object type is also provided, together with a list of complementary, or historical, object types inferred from the identifier list.

Objects can have *notes* to inform the user of some mistakes found in the literature,
or a link with some other object, or simply more extensive information connected
with this particular object.

Hierarchical links are proposed as links between a *parent* object and its *children*,
for instance, a double star and its components, a cluster of galaxies and the galaxies
it contains, a star cluster and its stars.

Similarly, *association links*, still under development, will specify the association of
two astronomical objects for any possible reason described in a paper.

- **Bibliographic references**: associated through citation in the paper of any of the
  object's identifiers. The bibliography gives access to abstracts and electronic articles when available either directly from publishers or through ADS[4]. Reciprocally,
  SIMBAD bibliography is systematically included in ADS.

- **Observational data**: also called *measurements*, they are always associated with a
  bibliographic reference. New entries have been recently added, namely distances
  and star diameters.

  Beside, external links are provided for some identifiers. Currently SIMBAD has a
  link to HEASARC[5] triggered by the existence of an identifier from a High Energy
  catalog and a link to some catalogs in VizieR (such as IRAS, 2MASS, HD, GSC,
  Hipparcos) and a link to the NED[6] database.

4.1.1   Identifiers

The identifiers constitute the basis for the SIMBAD name resolver facility that provides,
in response to any object name, the coordinates corresponding to the object position or the
list of papers citing the object. The name-resolving power of SIMBAD is used by many
archives and information systems (such as the archives of Hubble Space Telescope or
European Southern Observatory, the High Energy Astrophysics Science Archive Center,
the Astrophysics Data System, servers of the Digitized Sky Surveys).

4.1.2   Basic data

- Object types: from star to maser source, or cluster of galaxies, some 180 different
  categories, general, or very specific, are proposed. The object type list refers to
  a hierarchical classification of the objects in SIMBAD derived by the CDS team
  (Ochsenbein & Dubois 1992). It is evolving by adding new object types when required by the evolution in astronomy. For instance, the development of the ultra
  luminous X-ray sources (ULX) research domain yielded the creation of a dedicated

---

[4]Astrophysics Data System: http://ads.harvard.edu
[5]High Energy Astrophysics Science Archive Research: http://heasarc.gsfc.nasa.gov
[6]Nasa/IPAC Extragalactic database: http://nedwww.ipac.caltech.edu/

object type in SIMBAD. A complete list is available online on the SIMBAD website.

The object types are also a powerful tool for data cross-checking and quality control. They are also used in an experimental project of building an ontology for checking SIMBAD objects consistency in the frame of the Virtual Observatory.

- Coordinates: they are stored in the International Celestial Reference System (ICRS, see Feissel & Mignard 1998) at epoch 2000.0, after the publication of the Hipparcos and Tycho catalogs.

- Magnitudes: SIMBAD allows storage of any number of magnitudes and fluxes in the Vega or the AB photometric system. Historically, SIMBAD has stored B and V magnitudes. Currently this list has been extended to 13 magnitudes: U, B, V, R, I, J, H, K, and SDSS magnitudes u, g, r, i, z. They are provided with the associated uncertainty and bibliographic reference.

4.1.3 Bibliographical data

The creation of a list of papers citing an astronomical object requires a code to refer to each publication. A collaboration between SIMBAD and NED has led to the definition of the so-called *bibcode* assigned to every paper. This 19-digit bibcode contains enough information to locate the article (including year of publication, journal, volume, and page number) and has been rapidly adopted by journals and ADS. With the development of electronic publications, the page number is becoming obsolete, so too the bibcode. Publishers have defined a generic code, the digital object identifier (DOI), which is more adapted to current publication processes. SIMBAD will obviously follow this natural evolution.

The access to the list of references for any object is one of SIMBAD major products for users, but for many objects, there are many references where it may be difficult to find those of interest. To increase its usefulness, a flag is being added to represents the relevance of the citation in terms of astronomical contents for each paper linked to an object, so different relevance levels are possible: paper entirely devoted to an the object, object cited in the title, object in the abstract, object cited in a table, or object simply cited in the paper. This new flag is being tested and will likely be added for most new papers.

Several types of comments can be associated with the references in SIMBAD and are displayed after the reference:

- General comments: they are often comments added by the bibliographers about the problems encountered while cross-identifying the objects mentioned in the paper, typos in object names, etc.

- Notes about the existence of associated electronic tables or abstracts in the CDS server. Papers including no object are also flagged.

- Information on how the quoted objects are named in SIMBAD (comments related to the Dictionary of Nomenclature of Celestial Objects).

4.1.4   Some statistics on the data contents

The astronomical content of SIMBAD results from adding a selection into the database of *important* catalogs and published tables and a survey of the complete astronomical literature. This can be illustrated by a few histograms of the *V* magnitudes (Fig. 5) and of the 13 magnitudes already available in SIMBAD (Fig. 6). The list of the 30 catalogs for which SIMBAD has the most identifiers is displayed in Table 1.

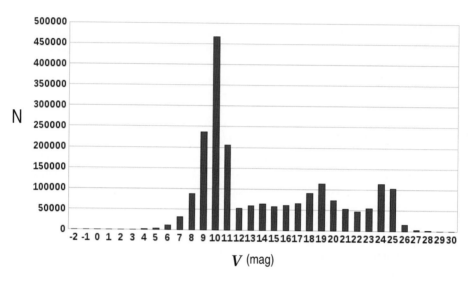

Fig. 5. *V* magnitude histogram. SIMBAD is complete up to *V* ≈ 10 mag (since inclusing of the Tycho catalog) and also contains very faint objects from the deepest fields observed so far.

Table 1. Main catalogs in SIMBAD with the corresponding number of identifiers in the database. A single source generally has several identifiers from different catalogs.

2MASS	1372142	FIRST	247749	HIC	118205
GSC	988949	2MASX	240411	HIP	118161
TYC	947069	CPD	231232	LCRS	92991
COSMOS	597301	CD	212186	CPC	86794
PPM	468709	NVSS	197587	[VV2006]	85219
Cl*	381965	LEDA	195290	GEN#	82344
HD	343325	AG	183566	UBV	73067
BD	319291	SDSS	162665	USNO	72798
IRAS	301189	YZ	137843	MGC	63309
SAO	258986	1RXS	124790	LSPM	61928

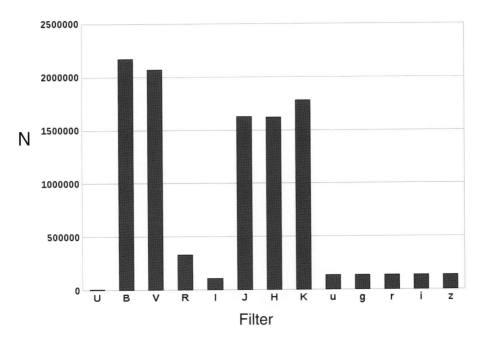

Fig. 6. Histogram of all available magnitudes. Historically only *B* and *V* magnitudes have been stored in SIMBAD but this is now extended to more passbands. The high proportion of sources with *JHK* magnitudes comes from a dedicated operation of massive cross-identification between SIMBAD and 2MASS performed at the CDS, which added the near-infrared magnitudes for ~1.6 million sources.

4.2 SIMBAD access

SIMBAD query modes are described in detail in the Appendix, to show how the wealth of available information can be accessed by users by simple to advanced queries, depending on their needs. The main access mode is the web, on which each query mode has its own form. Apart from the regular use of filling out a query form and getting the result in a webpage, the user has the possibility of submiting input files and to get the result in an ASCII file for further processing. SIMBAD can be accessed in two different ways: defining an URL with all requested parameters and getting the output as an HTML page or an ASCII formatted result; or using the SIMBAD defined web services.

5 Aladin

A web query to VizieR or SIMBAD produces a page containing information about an object or a list of objects. An interesting alternative is to investigate sky positions and to display the results of several queries simultaneously, superimposed on images. A few details on Aladin are given here, as an illustration of the possible use of data from VizieR and SIMBAD in a software tool.

Aladin (see Fig. 7) is an interactive software sky atlas allowing the user to visualize digitized astronomical images, superimpose entries from astronomical catalogs or databases, and interactively access related data and information from the SIMBAD database, the VizieR service and other archives for all known sources in the field. Created in 1999, Aladin has become a widely-used Virtual Observatory portal capable of addressing challenges, such as locating data of interest, accessing and exploring distributed datasets, and visualizing multi-wavelength data.

Fig. 7. Aladin: optical and HST images of the Trifid nebula. White circles are Chandra X-ray sources (Rho *et al.* 2004) with radius proportional to counts; squares are selected 2MASS sources for which an histogram of the $H - K_s$ color is displayed in the lower-right hand corner.

Aladin can be launched as an applet running in a web browser or as a standalone application. Installation instructions and documentation are available at `http://aladin.u-strasbg.fr/`

5.1   Accessing data from Aladin

Aladin gives access to a wide range of data services that delivers images or tables:

- Access to images: the Aladin software can access and query the Aladin image server, a dedicated database running at the CDS and providing optical images coming from the Digitized Sky Survey (DSS), as well as infrared images from 2MASS. Aladin can also load images coming from external services and archives, such as SkyView, the Sloan Digital Sky Survey (SDSS), the Multimission Archive at STScI (MAST) or the NRAO VLA Sky Survey (NVSS), or even local images over 1 GB. For most services, comprehensive metadata including the coverage and the orientation of the image fields help the user choose data of interest.

- Access to tabular data: for a given region of the sky, Aladin can retrieve information coming from the SIMBAD and NED astronomical databases. It can also query any of the ~8000 tables with positions available in VizieR, including large surveys such as NOMAD, 2MASS, USNO-B1, observation logs from missions such as Chandra, HST or projects!XMM-Newton, but also tables of individual objects published in journals.

In addition to the predefined list of available servers and to the access to VO-compliant resources, the user can describe additional servers in a configuration file, assuming they are available through HTTP calls. Moreover, local data in the form of FITS images, FITS tables, or tabulated format can be loaded into Aladin and compared with data retrieved from servers.

5.2   Visualizing and comparing data with Aladin

Aladin offers a whole set of tools allowing easy comparison between data covering the same region on the sky.

- Superimposition: once loaded, catalog data can be superimposed on an image of the same field. The user can then zoom in and out or pan the image by dragging the mouse. When selecting a set of objects, the associated measurements are shown in a dedicated panel.

- Combination of images: images can be combined to create a color *RGB* image or an animated blink sequence. The latter proves to be very useful when looking for objects with high proper motion or for transient events.

- Multi-view: the main window can be splitted into several panels allowing easy comparison of different images of the same region. The different views can be synchronized to show the exact same part of the sky at any time.

- Cross-identification: a cross-match tool lets the user cross-identify catalog data coming from different services, using the separation between the sources.

- Visualization of catalog sources: a tool allows filtering out and modifying the way catalog data are displayed (size, color), using information stored in the loaded catalogs. It allows, for instance, stars to be displayed as circles whose radius is proportional to the value of the flux or to display their proper motions as an arrow whose direction and length reflect the actual values of the corresponding parameters.

6   Conclusion

The CDS has constantly evolved since its creation, taking advantage in particular of the development of the Web, of electronic access to publications, and of course of the technical progress in storage capacity and network bandwidth, among others. The primary goals are to construct content from many information sources, especially papers published in journals, and to make it available to the whole astronomical community.

Metadata and normalization are absolutely critical ensuring easy access to data since they make it possible to select published data relevant to a project without having to read every paper. It also allows astronomers to find appropriate data among the increasing amount of available information (data mining). Such an ambitious goal implies a continuous concern for quality, which can only be achieved with the help of authors **and** referees. Special attention is given to providing data access simultaneously to non specialists and to providing detailed information that may be of use to specialists.

The content of VizieR and SIMBAD obviously reflects the literature, so references are systematically associated to the information stored in these databases. When errors or outdated information are detected by a user, it is advised to inform CDS, using question@simbad.u-strasbg.fr, and to provide as many details as possible with a reference so that a correction can be applied.

CDS has been at the forefront of networking online information, in close collaboration with the journals, the ADS, NED, and observatory archives. Thanks to the Virtual Observatory framework, CDS services, hence their selection of published information, are integrated better and better with other online resources, *e.g.*, through the Aladin portal, and through all other VO-enabled tools.

Appendix

A   SIMBAD query modes

SIMBAD has several query modes, using different kinds of criteria and allowing different outputs, both for direct reading and use in applications. Also some Virtual Observatory enabling facilities have been implemented.

A.1   Basic query

This is the simplest query mode. The form contains a single field that can be filled with an identifier, coordinates, or a bibcode, and it returns an astronomical object or a list of objects or the reference summary. This query mode can be added to the list of query servers in any browser's tool bar.

A.2 Simple query

A.2.1 Query by identifier

Any identifier understandable by SIMBAD can be submitted to the database to retrieve information known for this object. An option allows a query around the given name, also specifying the search radius. Query with partially known identifiers is possible by embedding wildcards (*=any string, ?=one char, [c1-c2]=one char from c1 to c2, ^ as first character in the brackets means one char excluding those specified). It is also possible to submit a list of identifiers prepared in an ASCII file, one identifier per line. The result can be a simple list of all the objects or a detailed display for each object found.

A.2.2 Query by coordinates

Queries by coordinates allow to retrieve all the objects contained in a circle defined by its center and a radius, which is 2 arcmin by default, and is limited to 10 degrees. A list of coordinates, each of which can be associated with a particular radius, can also be submitted from a user's local file. The result will contain a set of lists corresponding to each submitted position.

A.2.3 Query of bibliographic references

There are several ways to query the bibliographic references contained in SIMBAD. The basic way is to query a bibcode to get either the paper summary (title, authors, comments, ...) or the list of objects contained in the paper. The bibcode can be written using its regular syntax, with the possibility of using wildcards, or by specifying the different parts of the bibcode: year, journal abbreviation, volume, and page. Any field can be omitted. A reference summary can also be obtained by using several criteria, namely journal abbreviations, publication year range, authors, and words in the title.

  Any list of references obtained by such queries can be sent to the ADS for further querying, with the benefit of all ADS facilities.

A.3 Advanced query

A.3.1 Query by criteria

Queries by criteria allow to create samples of astronomical objects that share several common properties. SIMBAD offers more than 230 different criteria, such as basic data fields, existence of identifiers or measurement catalogs, bibliography in a range of years, and fields in the measurements. It is also possible to select the objects in a sky region, which can be a circle, an ellipse, a rectangle, a declination zone, a box defined by great circles, or a polygon.

  Most criteria can be defined by a value, a range of values, or memberships in a list of values. For spectral types and object types, there are two possibilities: either a strict correspondence with the given value (sptype = 'F3' or otype = 'star') or with all the underlying types (sptypes = 'F3' will include F3 type with all the luminosity classes and/or peculiarities; `otypes = 'star'` will include all kinds of subtypes of star, *e.g.*, variable stars, pulsars, carbon stars.

Query example: get all galaxies in a 30 arcmin circle having a *B* magnitude less than 16, and belonging to either the UGC or the MCG catalog:

```
otypes = 'gal' & region(CIRCLE,12 30 +10 10, 30m) & magB < 16
        & (cat = 'UGC' | cat = 'MCG')
```

A.3.2   Query by scripts

A SIMBAD script is made up of a set of commands defining the output, some common parameters like a default frame, and a radius for coordinate queries and any mixture of identifiers, coordinates, bibcode, and samples to be queried.

The output is always a straight ASCII file, with no embedded HTML tag. By defining its own output format (or a VOTable output with the needed fields), the user can easily ingest the SIMBAD output in its own application, maintaining a fair independence of upgrades made in the SIMBAD server software.

Example of a script, displaying, for a list of stars, an identifier (Hipparcos if it exists, HD if not, or any first one if HD is also missing), coordinates, B and V magnitudes, spectral type, and the number of references between 2000 and 2009:

```
format object f1 "%-10IDLIST(HIP,HD,1) | %-27COO(A D) |
       %FLUXLIST(B)[%4.2*(F)] | %FLUXLIST(V)[%4.2*(F)] |
       %-5SP(S) | %#BIBCODELIST(2000-2009)"
hd 1
hd 2
hd 3
```

The output of this query is

```
HIP    422 | 00 05 08.8331 +67 50 24.013 | 8.73 | 7.43 | K0    | 2
HD       2 | 00 05 05.950 +57 46 13.22   | 8.67 | 8.20 | F5    | 0
HIP    424 | 00 05 09.7582 +45 13 44.505 | 6.75 | 6.69 | A1Vn  | 7
```

Such a script can also be submitted from a user's file, which can be prepared manually or as the result of some application.

A.4   Output options

A dedicated form allows defining many output parameters, such as the kind of output that can be an HTML page, an ASCII output, or a VOTable, a list of data to be displayed or not, selection of a preferred identifier, range of reference years, notes. Most of the output options can be defined separately for full object display and lists of object display.

The output options are kept in a cookie and are reloaded during each session run from the same workstation and the same browser.

References

Bonnarel, F., Fernique, P., Bienaymé, O., *et al.*, 2000, A&AS, 143, 33

Cayrel, R., Jung, J., & Valbousquet, A., 1974, Bull. Inf. CDS, 6, 24

Feigelson, E.D., & Lawson, W.A., 2004, ApJ, 614, 267

Feissel, M., & Mignard, F., 1998, A&A, 331, L33

Genova, F., Egret, D., Bienaymé, O., *et al.*, 2000, A&AS, 143, 1

Genova, F., 2007, in ADASS XVI, ed. Shaw R.A., Hill F. & Bell D.J., ASP Conf. Ser., 376, 145

Jung, J., & Bischoff, M., 1971, Bull. Inf. CDS, 2, 8

Jung, J., & Ochsenbein, F., 1971, Bull. Inf. CDS, 3, 6

Lortet, M.-P., Borde, S., & Ochsenbein, F., 1994, A&AS, 107, 193

Lynga, G., & Westerlund, B. E., 1963, MNRAS, 127, 31L

Ochsenbein, F., 1994, Bull. Inf. CDS, 44, 19

Ochsenbein, F., Bauer, P., & Marcout, J., 2000, A&AS, 143, 23

Ochsenbein, F., & Dubois, P., 1994, in Astronomy from Large Databases, II, ed. Heck A. & Murtagh F., 405

Ortiz, P.F., Ochsenbein, F., Wicenec, A., Albrecht, M., 1999, ASPC, 172, 3790

Rho, J., Ramirez, S.V., Corcoran, M.F., *et al.*, 2004, ApJ, 607, 904

Wenger, M., Ochsenbein, F., Egret, D., *et al.*, 2000, A&AS, 143, 9

Subject Index

Name Index

Author Index

EAS Publications Series

Order directly to EDP Sciences

17 av. du Hoggar • BP. 112 • P.A. de Courtabœuf • 91944 Les Ulis Cedex A • France
Tel.: 33 (0)1 69 18 75 75 • Fax: 33 (0)1 69 86 06 78 • books@edpsciences.org

IAU SYMPOSIUM 284:

The Spectral Energy Distribution of Galaxies

SED 2011

5 - 9 September 2011, UCLan, Preston, UK

Topics :

- Stellar, interstellar and accreting sources of photons
- Properties of dust and gas ; radiation transfer
- Determination of star-formation rates and attenuation of light
- Panchromatic view of the Milky Way from radio to γ-rays
- Linking low- and high-energy properties of galaxies
- The cosmological evolution of emergent SEDs from distant galaxies

Scientific Organizing Committee:

Cristina C. Popescu (Univ. Central Lancashire, UK), co-chair
Richard J. Tuffs (MPIK,Heidelberg, Germany), co-chair
Gustavo Bruzal (CIDA, Venezuela)
Francoise Combes (LERMA, Paris, France)
Andy Fabian (Univ. Cambridge, UK)
Jay Gallagher (Univ. Wisconsin, USA)
Yu Gao (Purple Mountain Observatory, China)
Hidehiro Kaneda (Nagoya University, Japan)

Renee Kraan-Korteweg (Univ. Cape Town, South Africa)
Nick Kylafis (Univ. Crete, Heraklion, Greece)
Carol Lonsdale (NRAO, USA)
Vladimir Ptuskin (IZMIRAN, Moscow, Russia)
Elaine Sadler (Univ. Sydney, Australia)
Laura Silva (INAF, Trieste, Italy)
Jacqueline van Gorkom (Columbia University, USA)
Barbara Whitney (Space Science Institute, Boulder, USA)

Further Information : http://www.mpi-hd.mpg.de/sed2011

Picture credits: M82: X-ray:NASA/CXC/JHU/D.Strickland ;Optical:NASA/ESA/STScI/AURA/The Hubble and by Spitzer in the mid-infrared(red) Heritage team ;IR:NASA/JPL-Caltech/University of AZ/C. Engelbracht
Lake District Lou Johnson - Walking Britain http://www.walkingbritain.co.uk
Galactic Plane: ESA and the SPIRE & PACS consortia

SEPTEMBER 19-23, 2011 • OXFORD, UNITED KINGDOM

NEW HORIZONS
IN TIME DOMAIN ASTRONOMY

IAU Symposium 285

Invited Speakers

Brian Warner, South Africa (Keynote)
Sir Martin Rees, UK (Public lecture)

Suzanne Aigrain, UK
Isabelle Baraffe, Switzerland
Lars Bildsten, USA
Joshua Bloom, USA
Erin Bonning, USA
Phil Charles, South Africa
Jim Cordes, USA
George Djorgovski, USA
Laurent Eyer, Switzerland
Rob Fender, UK
Neil Gehrels, USA
Roger Griffin, UK
Josh Grindlay, USA
Franz Kerschbaum, Austria
Hans Kjeldsen, Denmark

Michael Kramer, Germany
Shri Kulkarni, USA
Don Kurtz, UK
Rachel Osten, USA
Stephen Potter, South Africa
Francesca Primas, Germany
Adam Riess, USA
Brian Schmidt, Australia
Alex Schwarzenberg-Czerny, Poland
Stephen Smartt, UK
Ben Stappers, UK
Mark Walker, Australia
Lucianne Walkowicz, USA
Nicholas White, USA
Rosemary Wyse, USA

Science Organizing Committee

Elizabeth Griffin, Canada, (Co-chair)
Robert Hanisch, USA (Co-chair)
Conny Aerts, Belgium
Dipankar Bhattacharya, India
Jianning Fu, China
Arne Henden, USA
Keith Horne, UK
Aris Karastergiou, UK/Greece
Katrien Kolenberg, Austria
Dante Minniti, Chile
Guy Monnet, France
Tara Murphy, Australia
Masatoshi Ohishi, Japan
Rob Seaman, USA
Alicia Soderberg, USA
Mark Sullivan, UK
Patricia Whitelock, South Africa

Local Organizing Committee

Aris Karastergiou (Co-chair)
Mark Sullivan (Co-chair)
Sarah Blake
Tom Evans
Vanessa Ferraro-Wood
Ian Heywood
Kate Maguire
Amy McQuillan
Yen-Chen Pan
Kimon Zagkouris

UNIVERSITY OF OXFORD

"Time discovers truth" - Seneca

WWW.PHYSICS.OX.AC.UK/IAUS285

Astronomy & Astrophysics

Aiming High

A weekly international journal which publishes papers on all aspects of astronomy and astrophysics (theoretical, observational, and instrumental) independently of the techniques used to obtain the results.

www.aanda.org

An international Publications series

Proceedings of international colloquia, workshops and schools, in the field of astronomy & astrophysics

European Astronomical Society Publications Series

www.eas-journal.org